PRAISE FOR

We Were There

"Latty's attempt to elevate blacks in the military beyond mere footnotes in history is timely and laudable."

—Columbus Dispatch

"There has always been a special poignancy about the military service of African Americans, especially during the long, shameful years of second-class citizenship. Yvonne Latty has collected scores of stories from black veterans about their experiences fighting for our country going back to World War II, and Ron Tarver's powerful photographic portraits put a face with every story. Most of them are proud, some of them are bitter, but their commitment to their country is an inspiration to us all."

—Mark Bowden, author of *Black Hawk Down*

"Eye-opening history, in the tradition of *Bloods,* the bestselling oral history of black Vietnam veterans. The veterans' are brief, but they accomplish more than many history books have dared even try."

—Detroit Free Press

"Yvonne Latty deserves a medal herself for this phenomenal book. Along with Ron Tarver's sensational photographs, *We Were There* vividly portrays the voices and personalities of incredible men and women who have risked their lives for this nation. These are stories of drama, courage, humor, and inspiration, and they have not received the attention they deserved—until now. *We Were There* is more than a work of history; it is a work of art, and one that will serve as an enduring tribute to these unsung heroes."

—Andrew Carroll, editor of the *New York Times* bestseller *War Letters: Extraordinary Correspondence from American Wars*

"An excellent introduction to combat experience. . . . It puts the African American experience in the broader context of American military history."

—*Booklist*

"Thanks to Latty and Tarver for introducing us to men and women who staked their lives on our freedom and security, but were never adequately thanked themselves. As we take in their images and stories our minds become living monuments to them, and our humanity and nation are made stronger."

—Lorene Cary, author of *Black Ice* and *The Price of a Child*

"*We Were There* is a tribute to all veterans."

—*Philadelphia Inquirer*

YVONNE LATTY is a native of New York City. She earned a B.F.A. and a master's degree in journalism from New York University before becoming a reporter for the *Philadelphia Daily News* and an adjunct professor of journalism at Villanova University. Ms. Latty has also freelanced with various Internet sites, including BET.com. She lives in Philadelphia, Pennsylvania.

RON TARVER has been a photographer with the *Philadelphia Inquirer* for over twenty years. His previous work includes a photo essay on African American cowboys, which is featured on the National Geographic Web site. He lives in Philadelphia, Pennsylvania.

We Were There

We
Were
There

Voices of African American Veterans,
from World War II to the War in Iraq

Yvonne Latty

WITH PHOTOGRAPHS BY

Ron Tarver

 Amistad

An Imprint of HarperCollins*Publishers*

A hardcover edition of this book was published in 2004 by
Amistad, an imprint of HarperCollins Publishers.

Grateful acknowledgment is made to the United States
government for permission to reprint the photographs of
Brig. Gen. Vincent K. Brooks on pages 176 and 177.

FIRST AMISTAD PAPERBACK EDITION 2005

Designed by Amy Hill

Printed on acid-free paper

The Library of Congress has cataloged the hardcover
edition as follows:
We were there: voices of African American veterans, from
World War II to the war in Iraq /[compiled by] Yvonne
Latty and [photographed by] Ron Tarver.—1st ed.
p. cm
ISBN 0-06-054217-9 (acid-free paper)
1. African American soldiers—Biography. 2. African
American sailors—Biography. 3. African American
veterans—Biography. 4. United States—Armed Forces—
African Americans—History—20th century. 5. United
States—Armed Forces—Biography. 6. United States—
History, Military—20th century—Sources. I. Latty, Yvonne.
II. Tarver, Ron.
E185.63.W4 2004
355'.0092'396073—dc22   2003058337

ISBN 0-06-075159-2 (pbk.)

Contents

Part Two: Korea

Part Three: Vietnam

Part Four: Persian Gulf War

Part Five: War on Terror

If there is no struggle, there is no progress.

—FREDERICK DOUGLASS

Preface

This is not only a book about war.

It is also about identity, growth, love, fear, bravery, and people who did more than they thought they could in the most difficult of circumstances. These twenty-eight men and women served from World War II to the War on Terror. Their ages at the time of service range from seventeen to forty-four. Some went on to great success, others struggled, forever scarred by their war experiences. In these pages, the veterans tell their stories in their own voices.

When I was a kid growing up on the border of Harlem, my father's service in World War II was one of his favorite dinner topics. I didn't understand how he could serve a country that discriminated against him because he was black, so I tuned him out. Eventually, he stopped talking about it, and I grew up never understanding the importance of what he'd done.

Years later, I was at my desk at the *Philadelphia Daily News,* where I am a reporter, when the phone rang. It was a black Vietnam veteran, Doug Culbreth, who asked me to write an obituary about his hero, George Ingram, a World War II Navy steward who served on a submarine. George battled terrible racism. He was skilled in many complicated jobs on the sub, but because he was black he was limited to serving as a steward. He washed dishes, served the officers their food, and did their laundry.

One rainy night in the South China Seas, a fellow crewman was high on the lookout tower while the sub was above water. The crewman, who was white, was trying to make his way down, but strong winds and thirty-foot waves kept crashing into the submarine and the crewman could barely hold on. Ingram inched his way across the deck in the blinding rain, and as a monster wave was about to crash down on the crewman, Ingram covered him with his body and held him to the rail, saving his life.

It took George fifty years to talk about what happened that night, or about any of his war experiences. The racism and indignities that were heaped on him by his commanders were something he wanted to forget. Not even his own family knew of his hardships or heroism.

I wrote George's obituary a few weeks after September 11, 2001, a time when I was feeling overwhelming sadness. I wanted desperately to feel a patriotic connection with my country, a connection I'd never had. Writing about George taught me that everyday African Americans played an important role in creating America. I realized there was living history all around me. I decided to seek out African American veterans and give them a forum in which to tell their stories.

Ron Tarver is a photojournalist for the *Philadelphia Inquirer,* which shares the same office building as the *Philadelphia Daily News.* Our paths had crossed on stories, in the cafeteria, and on the elevator, but except for a nod and a smile Ron and I were barely even acquaintances. However, I was a big fan of Ron's portraits and landscapes, which he exhibits nationally, and I knew that we would be perfect companions in this journey. Ron's task was to capture with a single image who the veterans are today. Each image had to tell a story. He spent a year traveling around the country, spending endless hours following the veterans through their daily routines, listening to them and photographing them.

To find these people, I used the Internet, newspapers, and veterans groups. In their stories I witnessed the progress toward racial equality. Many opportunities that I and other African Americans enjoy are a result of what these men and women endured. Their experiences are beyond the imagination of anyone who hasn't served in the military.

We have come a long way from the days when Magaritte Gertrude Ivory-Bertram, one of the first fifty-six black World War II Army nurses, couldn't get a white nurse to help her care for wounded patients because they didn't want to work side by side with a black woman. It's hard to imagine that there was a time when black men were not even considered good enough to die on the front lines with their white compatriots. World War II had a lot of firsts. In addition to the Army nurses, there were those like Luther Smith. Luther was one of the legendary Tuskegee Airmen. He flew 133 combat missions before his plane was shot down and he became a prisoner of war. World War II also saw the first all-black tank unit, and the first black WACS to serve overseas. More than a million black soldiers served in World War II, but because of their race, not one received the Medal of Honor, the nation's highest military honor for bravery. Fifty years later President Bill Clinton awarded it to seven black men. Only one, Vernon Baker, was still alive.

During the Korean War, African Americans finally got their shot to serve alongside whites. President Harry Truman signed Executive Order 9981 in 1948, which desegregated the military. When the conflict broke out, there were about one hundred thousand blacks in the military; by the end of the conflict more than six hundred thousand had served. Still, the Korean conflict saw the last segregated unit, the 24th Infantry. Robert Yancey was a sergeant in that unit, one of the first to see action. Some of these men served unrelieved for over a year, wearing summer uniforms in temperatures often as low as twenty degrees below zero. They had poor equipment and poor training, yet men like Yancey gave it their all. Charles Armstrong was a second lieutenant with the 2nd Infantry, a mixed unit on the front lines, and fought in some of the conflict's bloodiest and most infamous battles. After it was over, both Yancey and Armstrong wanted to continue with the Army, and both ran smack into bigotry. The Army struggled to desegregate and find a role for black officers. Yancey continued to serve; Armstrong, who lost his foot in battle, left enraged over the way he was treated.

Vietnam was the longest war America was ever involved in, and the role of African Americans in it was just as complicated as the cause of this long, bloody conflict. African Americans represented about 11 percent of those who served and about 13 percent of those who were killed. As the Civil Rights movement grew in the states, soldiers like James Brantley found segregation thriving in Vietnam. Like Alfredo Alexander, the majority of blacks served in the infantry, on the front lines. They found themselves in villages where women and children would shoot at them. They lived in the jungle, drank water from dirty streams, and watched fellow soldiers die in front of them. Young men with bright futures like James Robbins served. But Robbins and many others like him suffered so much from the trauma of war that they were never able to fully recover.

There is no accurate count of how many women served in Vietnam, but it is believed to be about seventy-five hundred. Black women were among them. Some volunteered to go. In Elizabeth Allen's case, you couldn't stop her. Marie Rodgers is one of the few African American women who served in an operating room on the front lines in Vietnam. She was awarded the Bronze Star by President Lyndon Johnson.

African Americans made up 23 percent of the troops deployed in the Persian Gulf War, and 17 percent of the combat deaths. African Americans were on the front lines during the Persian Gulf War with the Marines, Army, Navy, and the National Guard.

Lester Outterbridge served in the National Guard, and has the emotional and physical scars to prove it. He suffers from Gulf War Syndrome, and battles an endless list of ailments that have plagued him since he came home.

Today many African American men and women in the military work away from the front lines, in administration, combat support, medical and dental care. But during the darkest days this country has recently known, African Americans were among its heroes. Air National Guard Maj. Anthony LaSure protected the skies over New York, Philadelphia, and Washington after the attack on the World Trade Center towers on September 11. Air Force Cap. Eric Mitchell was sent to Iraq two weeks after he was interviewed for this book. Prior to that, immediately following September 11, he was in Afghanistan for the first battle phase of the War on Terror.

In Operation Iraqi Freedom, the general on every television station every day, explaining to Americans what was going on in the battlefields and skies over Iraq, was a black man, Army Brig. Gen. Vincent Brooks.

Men and women such as those in this book are all around you. They are in your neighborhoods, your churches, your offices, and even sitting on the bus next to you. African Americans have served with honor in every war this country has fought. It's time to say thank you.

—YVONNE LATTY

PART ONE

World War II

Leonard Smith,

Army Tank Corporal, 1942–1945

Leonard Smith was a cocky teenager from Harlem when he enlisted in the Army and became a member of the 761st Battalion, the first all-black armored tank battalion to fight in a war. Smith and about eight hundred other young black men who served in the battalion were used as an Army experiment to see how blacks would perform in tanks.

The battalion, also known as the Black Panthers, was part of Gen. George Patton's Third Army. They fought for 183 days in six European countries. According to Army accounts, they killed over 6,200 enemy soldiers and wounded 650, and captured 15,818 prisoners in some of the war's most violent battles, including the Battle of the Bulge, the largest land battle in World War II. The success of this battalion paved the way for blacks to serve in other armored tank battalions. But despite their heroics on the battlefield, it took thirty-three years for the unit to receive a Presidential Citation, which was finally awarded by President Jimmy Carter.

I volunteered when I was seventeen. I was playing hooky in school and this cop saw that I was getting into trouble. He told me that I should go into the service so that I would stay out of trouble. I thought it was a good idea. I wanted to go into the Air Force but they didn't take colored folks and the recruiters suggested the Army. After basic training, one of the sergeants decided to have me shipped out to be part of the tank battalion which was the first all-black tank battalion. I had never even seen a tank.

When we started out we didn't have a specific job. Everyone had to learn every position. We did drills for two years and then, in 1944, we were shipped out to England.

After a two-day stop in England, we were sent to the beaches of Normandy. We got there two days after D day. There were bodies all in the water and on the beaches. The bodies were decomposing. You could smell them. There was a lot of equipment all over the beach that was destroyed and muddy.

Our first stop was the French countryside, and that's where Patton himself addressed us. He had those two little pearl handguns on. I remember he said, "I only have the best in my Army. Don't let me down and don't let your people down."

Then they told us we had a little hill to take, but that wasn't true. They wanted us to take an area that a whole armored division couldn't take. We only had four companies* with five tanks, and all this firing was going on. The Germans were tearing us up.

In the midst of the fighting, my tank got stuck in a ditch and got filled with carbon monoxide. I passed out and the Germans thought I was dead. They pulled me out of the tank and left me hanging outside of the turret. For hours I lay there, and when I woke up I heard all these German voices talking. I pretended I was dead. I spent the night like that.

In the morning, the Germans finally left and I climbed out of the tank. I didn't know which way to go, and then I saw a German soldier coming toward me with a gun over his head, surrendering. Since I didn't have a weapon, I took his and told him to go. I followed him to the American line.

·   ·   ·

*Units consisting usually of a headquarters and two or more platoons.

The first two battles in France were the only ones that the 761st all fought together as a battalion; after that we were split up. My tank was paired with the 71st Infantry, an all-white unit. There were never any problems because we all had one goal: to get out of there alive.

One boy with the 71st and I were so close we were like twins. We were really good friends. But after the war was over I ran into him in a small town outside of Munich. First word out of his mouth was "nigger" and then he said if I were in his hometown he would hang me. My friends had to hold me back. I wanted to kill him.

During the Battle of the Bulge, my tank got hit and it wouldn't move. There were five of us in the tank and we couldn't get out through the top because snipers were shooting at us. So we decided to try and escape through a trapdoor on the bottom. As we started to crawl in the snow, the tank driver froze. He was so scared, he stood up and a shell came and took his head off. We crawled for about three thousand yards in the snow while the Germans fired at us. I wasn't scared: I was talking about the Savoy Ballroom and what I'd be doing if I were home in Harlem the whole time we were crawling.

They were really trying to kill us. They threw mortars, machine guns, everything they had. I wasn't scared, and that's what kept me alive. It was like playing cowboys and Indians to me. I never felt that I was going to die. We finally came to this ditch and went in, but it was filled with about fifty dead white American soldiers. I lay on top of this nice-looking soldier whose big blue eyes were wide open. We hid there for what seemed like hours. We weren't moving as long as the Germans were shooting.

One day in 1945, we were out looking for the Germans when the infantry told us there was some firing coming out of some kind of camp and they told us where to go. We came upon this big black-and-red gate with a barbed wire fence all around it. The gate was locked and there were barracks on the other side. We didn't know what it was. Another tank knocked the gate down and all these poor individuals came out of the barracks. You know when you go to a doctor's office and see a skeleton hanging on a string? That's what they were like, except they had a thin layer of skin. Some of them couldn't even make it to the tanks—they just fell to the ground. They were naked and their eyes were sunken in. They didn't say one word. They just stared at you.

When I saw them I went back inside the tank and grabbed all the cookies and chocolate bars we had. Then I got out of the tank and walked up to them. I touched them, hoping they would take the food. It was like touching bone. I tried to give them the cookies and candy but they wouldn't take it. I took a bite and then they took everything I had; they snatched it and shoved it in their mouths. In a matter of seconds I didn't have a thing. The tank commander told me to get back into the tank and that we were going to keep going, the infantry would secure the area. Later he told me that we were in a concentration camp that was a satellite of Dachau and that it held thousands of men.

Even though the people that were in the concentration camps say we were there, some people don't believe it,* and it makes me mad—these critics and whatnots going around saying that black soldiers were never in any concentration camps. Who are you to tell me where I was? I touched them and I fed them. We weren't assigned to go into a concentration camp—we ran into them by mistake. You want to confront these people who say you weren't there. You want to ask them, How do you know where I was? You have to have been with me to say I wasn't there. In Germany they acknowledge us, but here they don't want to give you credit for nothing.

I went back home on Christmas Eve of 1945. I decided to surprise my mother and so I just knocked on her door without her knowing that I was coming home. When she opened the door, it was not the same mother I left behind. She had aged a lot because she was so

---

*After the release of a 1992 PBS documentary, *Liberators: Fighting on Two Fronts in World War II*, which chronicled the experience of the 761st, critics of the films said that there was no way black soldiers could have liberated the camps. Camp survivors said the black tankers were there. Nonetheless, critics cited reports that showed the tanks were not in the vicinity at the time.

worried about me. Her hair was all white. I remember we held each other for a long time. Then I told her what I thought was good news. I had met a girl named Olga back in Munich and was going back. My mother started crying. I didn't want to break her heart again so I didn't go back. I stayed in New York and got my high school diploma. First I drove a gypsy cab, then I got city jobs. I was a sanitation worker, bus driver, and then a policeman.

In 1975, I retired to Fort Lauderdale, Florida, where I met a Buchenwald concentration camp survivor, Ben Bender, who became like a brother to me. He was freed by other members of my battalion, and is very outspoken about it. We went to Buchenwald together in 1991.

He took me through the whole thing of what happened to him. I cried when I saw the hooks they hung the prisoners by. I cried when I saw the furnace. After that trip Ben and I got very close. When he goes around to talk about being a concentration camp survivor I go with him and hold him up when he cries.

Margaritte Gertrude Ivory-Bertram,

Lieutenant, Army Nurse Corps,
1941–1945

Margaritte Gertrude Ivory-Bertram said her first day of nursing school back in 1937 was the beginning of the best years of her life. One of ten children born into a hardworking, churchgoing, affectionate Atlanta family, Bertram had the kind of confidence to take a chance on the Army. With the outbreak of World War II, the Army was desperate for more nurses and willing to give black nurses a chance. Bertram was one of the first fifty-six black nurses to serve in the Army.

By the end of the war, only 479 black nurses were serving in a corps of 50,000. Given the social mores of the day, the Army was reluctant to have too many black nurses tending wounded white soldiers.

I served as a nurse in the Army for four years, nine months, and thirteen days. I really wanted to serve, because it was an opportunity. I was always trying to improve myself. The Army offered more money, prestige, and a chance to travel. But I only got the chance because the war broke out. They needed nurses no matter what color, so they let us in.

The Army told me to buy a first-class train travel ticket to Fort Bragg in North Carolina, one of only two Army bases that took colored nurses. I was twenty-five and very excited, because I'd never traveled first-class before. But at the station things got crazy. I hadn't told the ticket agent that I was colored, and in those days colored people almost never traveled first-class. I made the reservation like they told me—first-class reservations to Fort Bragg—and they didn't ask my race.

A colored Pullman porter held up the train because I was on it. They didn't know where to put me. He was hostile. He said I was changing things and making trouble. "Girl," he said, "you are out of your place. You can't ride in the Pullman area."

In that moment I forgot all the great things that had happened to me since I became a nurse. All the respect I had received from the nursing school staff in my mind disappeared, and it hurt. But I snapped out of it and I walked over to the porter, looked him in the eye, and let him have it. I demanded my first-class traveling accommodations and reminded him that I was traveling under government orders. But he didn't care. "Girl, again I tell you that you got to be put in your place, and I am the one to do it!" It took them an hour to figure out what to do with me. He finally put me in a drawing room and said, "Stay in here and don't come out. I don't want you mixing."

He gave me a buzzer to ring if I needed anything. After a while I needed to go to the bathroom, and I rang it. The porter came in and pulled a green velvet drape off of something and it was a toilet; I couldn't even leave to go to the bathroom. When the train stopped, he would rush me out to whatever greasy spoon was at the station so I could buy something to eat. I wasn't allowed in the dining car.

When we finally got to Fort Bragg he said to me, "I am pleased to see that you are going into the service, but you better settle down or you will start another war about this traveling business. You young upstarts want to change everything overnight. It can't be done. Be patient and everything will improve." It was the first time I experienced prejudice directed at me by a colored person. I knew that colored porter went through a hard time keeping that job. I respected him even though I knew when he helped me off the train that he was glad to get rid of me.

When I got to Fort Bragg we were segregated. The dining room was mixed but we had to sit in the corner and then they later built us our own dining room. It was really no different than the life I was used to back home. I grew up in Clarkesville, Georgia, just outside of Atlanta, during the days of segregation. But I am from a strong family structure, and it helped me overcome these obstacles.

I'll never forget this one time at Fort Bragg when I was struggling up a steep hill in a night drill. I was sweaty and my legs could barely move over the rocks. I was stooping lower and lower under the weight of my backpack. Finally, I just gave out. Two white nurses saw me and ran over and helped me. Each one took an arm and they practically carried me to the top of the hill. When they got me to the top, the moonlight shined on my face and they saw my color. They roughly turned me loose, and I almost fell back down that hill.

I steadied myself and didn't say anything to them. I kept my temper, and felt sorry for them. Back then, if a white person was seen as being nice to us they would get criticized and ignored by their own people. In that way white people suffered too. They didn't want to make enemies among their friends.

From Fort Bragg they sent me to Liberia as part of the 25th Station Hospital, where we got a lot of burned pilots who had fought in Italy. We also took care of the troops who were protecting the airfields and rubber plantations. I was a supervisor.

It was 1943 and there were a lot of health problems in Africa, the worst being malaria and the lack of good drinking water. Even though we took precautions, we all got sick at one time or another. But a good thing came out of getting malaria: I met my husband. He was a staff sergeant and was visiting someone in the sick ward I was laid up in. I wanted to get to know him better so I gave a boy some chocolate to deliver a note to him. I didn't know what he would think but he answered me back right away. After I was discharged we had a few dates and he asked me to marry him.

. . .

Nine months after we got to Liberia, the nurses had to come back to the States because we had malaria. I was stationed at Camp Livingston in New Orleans, the only other base that accepted colored nurses. The mosquitoes there were bigger than in Africa and I was still feeling the malaria, so I asked for a transfer and was sent back to Fort Bragg.

The next year these flight nurses brought three planes of paraplegics wounded in the Battle of Normandy over to America, and the stopover was Fort Bragg. Forty-six young wounded white soldiers were all paralyzed from the waist down, and this was to be a short stop for them on their way to different Army hospitals across the country. The chief nurse told me to open a closed hospital ward for them. She said I should do everything I could to make them comfortable.

After the flight nurses brought them into the ward, the white nurse that was assigned to help me walked down the ramp, looked into the ward, turned, and walked away. I was busy and had no time to talk to her. I had patients in my care and no time to lose. Several corpsmen* were sent in to help me but the nurse never came back. I called again for help and was told the nurse didn't want to work with me because I was colored, so I worked without her. These men were paralyzed. Skin was coming off their legs and feet. I rubbed them with mineral oil, bathed them, put diapers on them, got a barber to fix their hair, and I made sure they had a good steak dinner. The soldiers didn't care what color I was. They just needed help.

I left in 1945 with the rank of first lieutenant. I wanted to get married and be a parent. I married my Army sweetheart, had a daughter, settled in Dayton, Ohio, and kept working as a nurse for over fifty years.

I should have received a commendation for what I did for those men in Fort Bragg. If I were white I would have. I'm not hostile

*Medics.

about the racism I went through. Though I didn't get what I deserved being a nurse in the Army, it helped me in my life. I was able to go back to school and further my education. We bought four different homes on the GI Bill. It helped me to learn about the customs of the people in the world. Christianity plays a great part in helping us to accept racism.

Luther H. Smith,

Air Force Captain, 1942–1947

On July 19, 1941, the Air Force began a program at the Tuskegee Institute in Alabama to train African Americans as military pilots. The men who trained in this program became the legendary Tuskegee Airmen, the first black military pilots.

A thousand or so African Americans served as Tuskegee Airmen, and 445 of them served in Europe, North Africa, and the Mediterranean during World War II. None of the bombers they escorted were lost to enemy fighters. They destroyed 251 enemy aircraft and won more than 850 medals. Sixty-six Tuskegee airmen were killed in action.

Luther Smith, who had wanted to fly since he was a boy in Iowa, became a Tuskegee Airman. He flew 133 combat missions and is credited with destroying two German enemy airplanes in the air and ten others in ground-strafing* missions.

My training at Tuskegee was the most intense period of training I ever experienced in my life. When I entered Tuskegee I was a licensed pilot; I could fly. I had taken flying lessons in Iowa, but the training for military aviation is as different as learning to drive an

* Low-flying aircraft firing machine guns at enemy military targets.

SUV compared to an 18-wheeler tractor-trailer. It was about precision, competence, and mission achievement. You do things the way you are told and you do it right the first time because your life is at stake. If you didn't, you were eliminated.

When we arrived, we were the Class of '43 and were to graduate nine months later. There were one hundred young black Americans. The commander, he said, "Look to your right and look to the left. When this class graduates there will be only twenty left." We were all college-trained, all physically sound and able to be military aviators. We were almost elite in terms of qualifications. When we looked at each other we couldn't imagine who wouldn't make it, but 80 percent of my class was washed out. You can't imagine the stress we were under on a daily basis. In May of 1943 when the twenty of us graduated, there was a bond between us stronger then a brother.

We were black. We had lived a life of racial prejudice, discrimination, and bigotry. We were used to be being considered second-class citizens, yet we had volunteered to join the military and fight in defense of the United States. We were educated, and we were just determined that we could become military aviators when white military leaders felt we lacked the courage and competence to do so. They didn't think we could fly sophisticated military equipment and maintain it. There was a lot of resistance. They accepted us only on an equal-but-segregated basis. We did not associate or train with or become a part of any unit of the Air Force. We were totally segregated.

When I got my wings, that was the biggest accomplishment at that point in my life. I had dropped out of college to go into Tuskegee.

In 1944, we went to Taranto in southern Italy and then were taken to Naples. I, along with others, had to protect Naples Harbor. When doing that, we never saw the enemy, but in April of 1944, Benjamin O. Davis* told us they needed escorts for heavy bombers. We would

---

* Commander of the 99th Pursuit Squad of the Tuskegee Airmen.

escort them from Italy to Germany and then bring them back to Hungary, where we would release them. We were also authorized to take out targets of opportunity, which were any ground targets we saw—railroad yards, air bases, and such. We called that "strafing."

After one mission, we strafed an air base in Budapest. I strafed two bombers. After we finished strafing the air base, the flight leader saw a freight yard and he decided to go down and do it. After he finished, we thought we were going in formation and going home, but when I looked at my wingman he went down to strafe the freight yard so I went down behind him. When I got halfway down, a massive explosion took place in front of me. I couldn't get away from it and was forced to fly through it. It was so massive it blew off the tail surface of my plane, and the rudders were pretty badly destroyed. The plane was intact but badly damaged. I was leaking fuel. I was burning a gallon a minute. I had two tanks of gas. When one was consumed I was going to switch tanks. I had one full tank and then there was another explosion. The coolant had leaked out, the engine got hot, and there I was over Yugoslavia with my engine on fire. I still thought I could get it back to friendly territory. I thought maybe I could use air and wind to cool the engine. My wingman said I was going to blow up. I had no speed.

I pulled out my radio jack and oxygen mask and then I proceeded to turn the airplane upside down so that I could parachute out. But the airplane was so badly damaged that when it turned, it went into a tailspin. I'm struggling. I got caught in the airplane and about that time my oxygen mask blew off my face and I thought this is how it's going to go. I thought it was the end. I went unconscious. Next thing, I was in my parachute looking at the airplane. It was burning up below me. I went unconscious again.

When I came to, I was coming down headfirst. At first I couldn't imagine what was going on, and then I realized what had happened to me. My parachute had pulled me from the plane while I was unconscious, snapping my right hip. My glove had been blown off and there was a bruise on my hand. I was in a state of shock. My right foot was turned backwards and was broken. I thought I was not going to make it, and again I went unconscious. When I came to, I was crashing through the trees.

The Germans saw the parachute in a tree. They climbed up the tree and lowered me from the tree by placing rope underneath my armpits. They saw that I was badly injured and could not walk. They put me on the back of a horse and carried me about a hundred yards and then put me in a car. They wrapped me in a parachute and took me to a local hospital. I got first-aid treatment. They set my leg on a splint and kept me overnight. Then they delivered me to a larger community hospital in Yugoslavia. I remained there for a couple of weeks and then they took me to Austria and placed me in Stalag 18, a German prison camp.

It was not like being at your favorite hotel. Life was extreme, but I was treated with respect. The German people are very militaristic. They recognize rank and they respect it. Since I was an officer, although I was black, they treated me how they treated their own people. Shortly after, I had a visitor who I believe was Swiss but was a representative of the Geneva convention* and was there to check on how they were treating prisoners of war. He said he was assigning me a prisoner of war number and I would be as safe as if I were at home. The person said that as an officer I was not supposed to work and do other things. He said they would treat me with respect but don't expect any luxuries.

In Germany, prisoners of war in October of 1944 were the most valuable commodities the German nation had. They were losing the war, and the most negotiable items they had at the end to negotiate peace terms would be their holding of Allied prisoners of war. So we were pretty important to Germany. There was no rough stuff, no mishandling at all. The other side of the coin was that they were losing the war and they didn't have much to work with. Food was something that almost didn't exist and there were days where the

---

*Signed in 1864, this treaty that guidelines for the humane treatment of POWs, civilians, and the wounded in wartime.

food I ate was grass soup, which was just grass and hot water—not very nourishing. So I began to suffer from malnutrition.

They kept insisting that I speak German and I'd say, "*Nein, nein.*" I thought the war would be over in a day or two. I deteriorated steadily—my hip was in two pieces. I was in the prison camp. They didn't have a hospital facility. So from late October, when I went in, until late November, I was expecting to die. My situation was deteriorating to the point that the doctors told the Germans that I was dying. In late November they transported me to a hospital in a little village in Austria. The little town was in a ski resort. The Germans had converted resort hotels into hospitals.

In December, they played Christmas carols, but on the sixteenth of December in 1944 they stopped playing carols and instead there were a tremendous amount of speeches. It so happened that Canadians working near the hospital asked to visit me, the American prisoner of war. I asked them what was all the stuff on the radio. They said it was news of the Battle of the Bulge. It was Hitler on the radio, and he was telling the German people that they were fortunate, that they had the enemy surrounded and now would have time to produce their secret weapon—which was the V-1, V-2 rocket—and use them against England.

This scared the daylights out of me. I had not wanted to learn German because I thought the war was going to be over any day. Suddenly after the speeches I felt I should learn German, and I did in three weeks. They had placed me in a room with another prisoner of war who was South African and white. He was a POW and spoke German. He didn't like the way I spoke German. When I spoke, he interrupted me and lectured me on how to speak correctly. In three weeks they were asking me if I had ever lived in Germany, I had learned so much. The war began to die down. FDR died and the Germans began asking me about Truman, who he was and what would happen.

While I was in the hospital, in walked a German SS officer. He wanted to carry on a conversation. He was very arrogant and the first thing he said was, "Your propaganda is the best in the world."

I said, "*Nein, nein.* Germany's is the best in world."

He said, "Your president has convinced you Americans that Russia is your ally. Russia is your enemy. I predict you will be at war with Russia in five years. Furthermore, your propaganda is so good it's duped you. You are black American." Then with utter contempt he said, "You volunteered to fight for a country that lynches your people."

Here this arrogant SOB was saying this to me and my first reaction was, you are absolutely correct. I told him that I didn't understand and he walked out of the room. I was floored—you might as well have hit me with a heavy stick. Yes, I had volunteered to fight for a country that lynched my people.

The next day he brought it up again.

I said, "You people are just as bad. You lynch your people. Your German Jews, you lynch them. I am black American. It is my home. I will fight for it because I have no other home, and by fighting for it I can make America better."

I told the Germans I wanted to be sent back to the POW camp with the other Allies. The Russians were advancing on where I was and I did not know how the Russians would treat Allied prisoners, so I wanted to get back to the camp. I went back to Stalag 18 in April.

When the Allied forces liberated the prison camp in May, they rode me up to a doctor on a gurney and he said, "Hi, how are you? The war is over." He asked me a couple of other things about my condition and about my wounds and my legs. But every time the doctor asked me a question, I would answer the question and the attendant would repeat my answer and then the doctor would write it down. I was responding in German without realizing what I was doing. The attendant was translating what I was saying to English. At one point the doctor patted me on the shoulder and said, "You'll get over it, soldier." German was the only language I could speak.

.   .   .

When I came home, I had operation after operation. I weighed only seventy pounds and didn't have much healing capacity. I had an operation that molded my hip together but because of atrophy, my right leg is seven inches shorter than my left. When I was stabilized, I went home for ninety days to live like a human being. Then I was hospitalized for the next two years.

I was awarded the Distinguished Flying Cross,* Air Medal with six Oak Leaf Clusters,† Purple Heart,‡ European and Mediterranean Theaters Campaign Ribbons,§ and the Prisoner of War POW Medal. After I left the Army I went back to school and got a degree in engineering from the University of Iowa. I worked at General Electric Company and was awarded two patents and published numerous technical documents and publications. I worked on special assignments with the United States Air Force, NASA, and the United States Navy Submarine Command. Now that I'm retired, I speak in schools about my experiences as a Tuskegee Airman. It has shaped who I am.

*Awarded to officers and warrant officers for an act or acts of valor, courage, or devotion to duty performed while flying in active operations against the enemy.

† Awarded to U.S. and civilian personnel for single acts of heroism or meritorious achievements while participating in aerial flight, and to foreign military personnel in actual combat in support of operations.

‡ Awarded to those who are wounded or killed in action.

§ Awarded to those who served in those regions.

Thomas Hayswood McPhatter,

Marine Sergeant, 1944–1946;
Navy Reserves Chaplain, 1953–1959;
Navy Chaplain Captain, 1959–1983

Iwo Jima is a tiny island, one-third the size of Manhattan, dominated by a 550 foot volcano. Only 660 miles south of Tokyo, it stood halfway between the Japanese and American bomber bases. It was also the home of three Japanese airfields. More than 110,000 U.S. Marines in 880 ships were sent to this island to fight one of the most ferocious battles in World War II history. The first black Marines were among them.

These Marines were not allowed on the front line, but on this island, everywhere they stood was a front line. The Japanese fought from underground caves, emerging only in suicide missions in which they swung swords and vowed to kill ten Americans before they were killed. In thirty-six days of fighting, one in three men was either killed or wounded.

Thomas Hayswood McPhatter, now of San Diego, was looking for an opportunity to serve his country when he became a Montford Point Marine and one of the first black Marines. He was a sergeant in the 8th Ammunition Company of the Marine Corps and was in the D day battle of Iwo Jima.

.   .   .

When we first got near Iwo Jima there were kamikaze planes crashing into ships and dogfights in the sky between the Americans and Japanese. The first three waves of Marines that went into Iwo Jima were crying on the radio for help. As soon as they hit the beach, the Japanese would kill them on shore. So they sent in more planes to bomb some more. Instead of going in on D day, we went in on D day plus one.

Iwo Jima was the most dismal place I had ever seen in my life. There were bodies on the beach. The sand was black with volcanic ash. There was no live foliage. If you dug more than two feet your hands would get hot because of the sulfur on the island. You didn't have any real foxholes. Quite often I tried to dig a foxhole and wound up digging up bones. The Japanese had been burying their people wherever they dropped. In Iwo Jima the smell of death was everywhere. You could smell it and taste it. Big green flies, as big as grasshoppers that shined like they wore armor, were eating the flesh of the dead. With our dead you had to take them down to a designated place where a makeshift cemetery would be. There were holes, and that was where you put the remains. There was never really a whole body. Then they'd put one layer of dirt on it and you'd put on another, like a cake. They'd put up these little crosses with dog tags, but no one really knew who was under the grave. An all-black unit took care of the bodies.

The only jobs we could have in the Marines were either taking care of the dead or ammunition, which is what I did. I joined the Marines because I thought I could avoid bigotry and racism, but I ran right smack into it. No matter what skills I had, all they would let me do was take care of the ammunition. My job was to bring the ammunition from the ship to shore and get it to the troops that were on the front lines. I'd also fuse the rockets and mortars, the heavy projectiles, and get them to the soldiers. We moved it in by DUKW,* a vehicle that goes on water and land. Sometimes it got

---

*From a sequence of codes used by General Motors Corporation : *D*, 1942 (first year of production) + *U*, utility truck, amphibious + *K*, front-wheel drive + *W*, tandem axel.

stuck on the black ash and wouldn't move and we would have to push, pull, or leave it there.

There was stuff going on all the time. You could hear it whistle by you, men getting hit right near you. You didn't know when yours was coming. As the troops moved up, we moved right behind them. At night you couldn't tell who was who. The Japanese would dress up like Marines. They would take the uniforms right off the dead Marines and they'd come right into your area and the only way to protect yourself was by using a password. If you didn't get the password right we'd start shooting.

The ammunition was on the center of the island, but at one point the enemy dropped mortars on it and all the ammunition went off. We had to run for the beach. We were hollering the password, smoke was all around us, and the island was shaking. I got smoke burns on my face from the explosions. We lost four men from my platoon that day.

When we went back, the Marines were screaming for hand grenades, rockets all the stuff that had got blown up. They decided then to have planes drop ammunition—using parachutes—and we were supposed to run to them, gather the ammo, and get it to the troops.

You never knew how or when the enemy was going to be shooting at you. A shell came at me one time and I had to run into a bunker. A white Marine was in there, dead. He had pictures of his family and a letter to his mother in his hand. I had to put my head into the sand to protect myself. With my head in that hole, I made a promise to God that if he let me live, I'd give him my life in service.

All the Japanese installations were underground, which is what made it so hazardous. Toward the end of the battle the Japanese put up one last fight. We had tents in what we thought was a secure area. We had dug out holes and put in what we called "rubber ladies" which is a flotation tube. One day they came up out of their holes. They knew they were going to die. They were swinging swords and screaming, "Banzai, banzai." They had this idea that they were going to kill before they were going to be killed. They cut the ropes so the tents would fall on the soldiers. Then they stabbed them. Someone fired a warning shot so we knew they were coming, but it wasn't until they hit our area that we were able to stop them.

I was in Iwo Jima for two and a half months. We were under the impression that we were going to be sent home because of combat fatigue. We were sent back to Hawaii for R&R but the next thing you know, we are being sent back to invade Japan. I was about two days away from Japan when they surrendered.

You know, the only time I was treated as a human being in the Marine Corps was by the Japanese after we conquered them. Less then a week after the bomb was dropped we pulled into the harbor of Hiroshima. The Japanese received me and treated me well, even through they were hungry. That was not my experience with the Marine Corps. I found the corps to be racist, particularly when I had to interface with authority. I wasn't expecting any kind of a payoff other than to be treated like a man, but even that was hard for them.

As long as I was in Montford Point in Jacksonville, North Carolina, where they trained black Marines, it was wonderful. Black men ran that place. It was the black West Point, and I went in the second year they allowed blacks to join the Marine Corps. You didn't have any black officers but it was certainly run by black veterans, of the 9th Cavalry and the Buffalo Soldiers. We had all white commissioned officers. I was a noncommissioned officer but I was a sergeant. I couldn't go any higher than a low-grade sergeant, even though I had more responsibilities. We didn't even have the same designation as the white Marines. We were called USMC Selective Service, which meant we were on trial.

While stationed in Hawaii, I wanted my mother to contact the NAACP or someone because they were treating black Marines so bad. I wrote a letter to tell her what was going on. All the letters were censored, and they read my mail. One day, instead of going out for training, I was held to see a commanding officer.

"Didn't you know we were going to read it?" he asked.

"Yes, sir. I couldn't tell you any other way or I would be charged with insubordination," I replied.

"What do you want me to do with it?" he asked.

I asked him to send it off to my mother. Shortly after that there was a mini-riot. Frustrated black soldiers were firing live ammo.

"You lads seem to want some action," the commander said. "I'm going to send you where you'll get more action than you planned for."

He sent us to Iwo Jima.

When I came back to the States after the war, I went back to Montford Point. I had enough points to be promoted but I didn't want to stay in the Marines. I never forgot that promise I made in the foxhole in Iwo Jima to give my life to God. And that's what I did.

Waverly B. Woodson Jr.,

Army Medic Corporal, 1942–1945;
Army Reserves Sergeant, 1945–1952

Dday was the turning point of World War II. The allies stormed the beaches of Normandy, France, in the largest seaborne attack ever. About 150,000 men took on Hitler's army with one goal—to drive the enemy back to Berlin. But the Allies were under relentless attack from the shore. Close to eleven thousand were killed or declared missing in action. Many never made it out of the water.

Waverly Woodson was an Army medic with the 320th Antiaircraft Barrage Balloon Unit, an all-black group that saw action on D day. Although on that day he was dubbed by the black media as "Hero No. 1," before long, Woodson said, he was forgotten. And most people don't even know that he and more than a thousand other black men were at D day.

If you ever want to know what hell is like, D day was it.
I was on what they call a landing craft, which is a ship that carries troops, vehicles, and supplies onto the shore. I was in the back of the first wave, and what I saw in front of me was death. All these men, thousands and thousands of men, were heading onto the beach, and the Germans were just shooting them down. Mortars

and shells were hitting boats. There were a lot of dead bodies floating in the water. On the other boats around me, you could see men
getting seasick one minute and the next minute they were in the
water, dead.

Our ship ran over a submerged mine and then the Germans
started shooting at us. Next thing you know a shell hits us on the
right side and it blew up all of our vehicles except a tank, and killed
fifty-five guys. The only ones that made it were me and two other
guys, but I was hit in the back and groin. What was left of our landing ship tank (LST) was just floating along. And then another shell
struck it and I was thrown in the water. There was a lot of debris
and men were drowning all around me. I swam to the shore and
crawled on the beach to a cliff out of the range of the machine guns
and snipers.

I was far from where I was supposed to be, but there wasn't any
other medic around here on Omaha Beach. It was just about ninety
minutes after the start of D day. I had pulled a tent roll out of the
water and so I set up a first-aid station. It was the only one on the
beach.

Bullets and shells were flying everywhere. It seemed like everybody was either dead or screaming, "Doc, Doc, help me." I took a
bullet from the shoulder of one, and dressed the gaping hole in the
shoulder of another. I even amputated a right foot.

At one point I heard all this shouting coming from the sea.
There were thirty Tommies [the nickname for British soldiers]
drowning, so I waded in and dragged out four of them and gave
them artificial respiration and revived them. Then I told other guys
what to do so they could save the other men.

They say I saved three hundred men but I couldn't tell you how
many. The newspaper articles I got here says I worked thirty hours
after being hit, but I can't remember. I just know it was a long time.
After saving those Tommies, I collapsed. I think I had lost too much
blood, even though I had wrapped up my wound.

They put me in the hospital for two days but I had to go back; I
wanted to go back, I don't know why, I just did. So I went back for

another twelve hours on D day plus four and worked until I was so exhausted I just couldn't do no more.

After the war I would have gone to Korea but I was in love with Joann and we had just got married. She didn't want me to go. So I worked at Walter Reed Army Hospital in Washington, D.C., as the sergeant in charge of the morgue. I did autopsies. We raised our family. I got two girls and a boy. We got lots of land here in Clarksburg, Maryland, and we're happy.

When you talk to white people they will swear up and down that there were no black people in the first wave, but I went in without my unit. I was put on another LST. I'm very light-skinned and a lot of the men thought I was white. Newspaper articles say that the War Department sat on my commendation. They finally gave a Medal of Honor to some black men who served in World War II, way after the fact.* I did get a Purple Heart† and a Bronze Star.‡ Joann, my wife, made a display case for it and we hung it up near the door.

On the fiftieth anniversary of D day the French government recognized me. It sent three of us on a weeklong, all-expenses paid trip to France, where they gave me a medal during a ceremony on Omaha Beach. I don't know why they chose me, but it was a wonderful thing. I was the only black man of the three. I think it was the French's way of saying, "Thanks."

*The Medal of Honor is the nation's highest military honor for bravery. Fifty years after the end of World War II, seven African Americans were awarded the medal by President Bill Clinton. Only one, Vernon Baker, was still alive.

† The Purple Heart is awarded to those who are wounded or killed in action.

‡ The Bronze Star is awarded to individuals who, while serving in any capacity with the armed forces in a combat theater, distinguish themselves through heroism, outstanding achievement, or by meritorious service not involving aerial flight.

Gladys O. Thomas-Anderson,

Private, Women's Army Corps, 1944–1946

With their heads held high, their backs erect, and regal expressions on their young faces, hundreds of black women walked in tight formations of three through the cobblestoned streets of Birmingham, England. These women, members of the 6888th Central Postal Directory Battalion of the Women's Army Corps, held a role of distinction. Because politicians and the black community loudly complained about discrimination in the Army, approximately eight hundred black enlisted women were sent overseas to sort backed-up mail. Besides nurses, members of the 6888th were the first and only black women sent to Europe during World War II.

During this march through the streets of Birmingham, the 6888th were dressed in their finest uniforms, which included spotless cream-colored gloves and berets they called "hobby hats." The women oozed class, even as their hearts pounded with angst. But as the battalion marched, their fears eased, because the loud cheers wouldn't stop. The parade-goers whistled and clapped. The streets were lined with a warm, enthusiastic all-white crowd.

As she marched on that warm day, Private First Class Gladys O. Thomas could not believe what she was seeing.

.   .   .

My brother and I always did things together. When the war was
going on in Europe, he knew they were going to draft him eventually so he went in because he was unemployed and had no
prospects. When he went, I wanted to go, too, but my mother said
no. I was only eighteen. But at twenty-one I decided to go in. I knew
this lady who was in the service and left because she had a bad
foot. We were in a bar one night and she was telling me and four
other girls about the Army and we all decided to go in, but we had
to take a physical. One was too fat, the other was too skinny, and
the other was borderline. That left the ex-WAAC* and myself, so I
signed up.

I was sent to Des Moines. Well, I've never been in on anything
like that before. We had to learn how to keep our clothes in a certain way—shirts in one place, skirts another, and jackets another;
everything had to be just so. We had to go to basic training every
morning and had to fall in line. Then we'd work on how to salute
and make our turns. I wanted to travel, and after training we were
all sent to different places. I was sent to an Army Air Force installa-
tion. I was supposed to be in the military, yet they made me pay my
own way to California.

We always tried to look good and took care of our appearance.
When I was stationed in California, a white man stopped us and
asked us why our uniforms looked better than the white girls'.
Well, we always made sure our uniforms were pressed and not
wrinkled. Shoot, a lot of the girls worked in laundries back home—
before the war, jobs were not that plentiful. They knew how to iron.
They could get their uniform together in five or ten minutes. Then I
heard that the Army had decided they wanted to send black WACS
overseas. Even though I wanted to go, they told me they weren't
going to send me, that they weren't taking any girls from my base.
But one day I went out to get my cleaning and my lieutenant said,

*The Women's Army Auxiliary Corps worked with the Army but was not considered part
of the Army. Later it were incorporated into its Army and the name was changed to
Women's Army Corps.

"Where the hell were you? Get ready—you are going overseas." That's how I became a part of the 6888th Battalion.

On February 3, 1945, we left for Scotland on the *Ile de France*—it was a luxury ship they outfitted to transport troops. It took about eight days to get there. I loved the boat ride, even though it was kind of hectic. There were a lot of heavy waves and one night I felt a big lurch and all our bags rolled across the floor. They told us they thought a German sub was in the water. But I didn't believe it and I just rolled over and went back to sleep. Turns out there was a sub chasing us, but it didn't get us. When we docked in Scotland there were bagpipers playing to greet us. From there we went to Birmingham, England.

We were treated better in Europe than in the States. They thought we were beautiful. A lot of them had never seen a black girl dressed in a uniform. They'd whistle at us all the time. You could go anywhere you wanted; there was nothing you couldn't do. We never saw signs that said Colored Only.

Every city in Europe we worked in, it was part of our routine to march in front of dignitaries, commanders, and such, and the crowds were always crazy about us, people just going wild for us. I believe we were the first black women they had ever seen. There were always hundreds of white folks lined up to watch us and let me tell you, it was a feeling like no other I'd had before. It was amazing, exciting, and it felt so good.

When we were on furlough in Switzerland, my friend and I went into a store and bought watches. The man said that he wanted to take us to a nightclub. I thought he was just talking about the girl who was with me—she was real tall and pretty; I was just average. But he wanted to take us both. We went out with him and had a great time. He took us to a place where the music was like Glenn Miller.

Sorting the mail took a lot of concentration and it was dull. We were just changing numbers and writing lists. We sat three to a table and there was an index-card box there. You'd pull out a card, check the name on the list, and change their address. We worked eight-hour shifts around the clock. We never had the same schedule. Just when you got used to one schedule, it would shift. But mail was important to the GIs, because it was something that came from home. It was a way they could find out what was going on. It was personal and raised their spirits. We did a bang-up job of it, too. The mail had been sitting in a warehouse for a long time before we sorted it all out.

In Birmingham, we slept in an old school. We took cold showers and our beds were on the floor. There were about ten of us to a room. Six months later we transferred to Rouen, France, near Paris, and my job didn't change at all.

Paris was magical. It was the most exciting thing to be in this group and be in Paris. You could drink out of any water fountain you wanted to. We went to lots of parties, dance halls, and had a lot of fun. I knew I was there to do a job, which I tried my best to do. But I also wanted to have some fun. I wasn't thinking so much about the war. I just wanted to see things, have experiences. I had never been out of Detroit.

After the war the battalion was split up and I was assigned to a unit in New York doing clerical work—filing and stuff—in a hospital. I was in Staten Island one day boarding the ferry and realized I was bored. I was in great big New York City and I was bored. I decided then that I'd had enough, and went back home to Detroit. Sometimes I wish that I had stayed longer, but my Army buddy and me were separated—they sent her to California—so I didn't really have any close friends, and you need to have someone you have a common thread with to be happy.

I heard that my battalion left Europe in 1946 with the distinction of having broken all records for redirecting mail. Our unit was supposed to be cited for this, but to this day I haven't heard anything to that effect. Being the low man on the totem pole, you get forgotten about.

.   .   .

I'm proud that I was able to serve. I'm proud that I was chosen. I was awarded the Victory Medal,* Good Conduct Medal,† and Army of Occupation Medal.‡ It was an honor to be in the Army and be black, too. We were the beginning. We were the ones who set the pattern. Now you have black pilots and so forth. It's a wonderful era. I've always been proud to be an American, and even more so after being in the service.

---

* Commemorates military service during World War II.

† Awarded for exemplary behavior, efficiency, and fidelity in active military service. It is awarded on a selective basis to each soldier who distinguishes himself or herself from fellow soldiers by exemplary conduct.

‡ The medal was awarded for thirty days' consecutive service while assigned to post–World War II Europe, Japan, and Korea.

James Tillman,

Army Sergeant, 1942–1945

James Tillman of Pittsburgh, was drafted a year after the attack on Pearl Harbor and became a "Buffalo Soldier" when he was just twenty-one. "Buffalo Soldiers" was the nickname of the 92nd Infantry, the only black infantry unit to fight in Europe during World War II. It was a nickname that dated back to the late 1860s when black soldiers volunteered to tame the West. They took on outlaws, American Indians, and Mexican revolutionaries. It's said that the American Indians nicknamed them because they thought that with their dark skin and curly hair, the soldiers resembled buffaloes.

The 92nd Infantry consisted of twelve thousand soldiers, which included two hundred white officers and six hundred black officers. Most of these men were poor Southerners. Many of them had never learned to read or write, but it didn't matter when it came to fighting for their country: They valiantly fought some of Hitler's toughest and best-trained German troops in the rugged mountains of Italy. When the fighting ended, 343 soldiers from the 92nd Infantry had been killed, 2,215 wounded, and 615 were missing in action.

At first I was a conscientious objector. My dad was a minister, a strict Christian. He had the paperwork all filled out for me, but when I got to the draft board they told me that if I was a conscientious objector I would get no pay, have to work on roads and things. I didn't like that idea so I changed my mind. They told me not to worry about being on the front. I was a truck driver. I'd just drive trucks for the Army; I wouldn't be in combat. At that time no black men were on the front, so I signed up.

But things weren't how they said it was going to be. They were making up a division, the 92nd. It was very big and the first all-black outfit. They couldn't put us all in one place because we were in the South and they didn't want so many black men together in one state, so they put us in four different places.

Even though we were supposed to be infantry, they didn't know what to do with us. From Maryland they sent us to Louisiana to train in the swamps. They were going to send us somewhere in the South Pacific, but while we were training, that campaign ended. Then they said they would train us for desert fighting and send us to Africa. They carried us to a desert camp in Huachuca, Arizona, and the whole 92nd Infantry was joined together. But then the Africa campaign ended, too. So now we had been training almost a year—most outfits train for three to six months—but they kept moving us around. They were talking about breaking us up, but politicians didn't want them to. Blacks in the Army usually took care of ammunition and stuff, but black politicians wanted us on the front lines to shed blood so we could get recognition; just being in the Army wasn't enough. The white man wanted the prestige of fighting on the front line. He said that since we had no rights and no privileges in our country, we wouldn't fight for it when the time came.

After a year and a half of training—the longest training of any unit, I believe—we still didn't know if we would get the chance to fight. But I was willing to do my part. I got caught up in it. The flag had no meaning to me before, but the flag was the only thing that was not segregated. The white soldiers had the flag in front of them and we did, too.

Finally, they had us train for the Italian mountains in Arizona and called us up when the Allies got up to Naples. When we got there, the Allies had already taken part of Italy and we were moved right up to the front. We were there before daybreak and relieved a white unit. That was the only time I came in contact with white troops.

We had a year and a half over there. I was a gunner and used the heavy weapons, the mortars. We took a lot of cities. From Naples we went all the way through Italy. We helped take the cities of Rome, Florence, Milan—every city we went to was destroyed. But the Germans didn't do it to Rome. The Pope asked Hitler not to destroy Rome and he didn't.

It was rough all the way, but we were dedicated. We were fighting for a greater cause, for our people. I didn't want to see what they were doing to the Jews happen to us, and the Germans wanted to do it to everybody. We had to defeat them and we had to prove that blacks would fight. We knew we could fight with anybody, and that is what drove us. We wouldn't quit. If we failed, the whole black race would fail. We were fighting for the flag and for our rights. We knew that this would be the beginning of breaking down segregation.

I lost quite a few of my men, lost as many men through mines as through action. When the War in Europe ended, I was on the front lines in the mountains near Genoa. We were getting ready for an attack when we heard the war was over. But we didn't move until we saw the Germans come off the mountain. There were thousands of them, and they left all their weapons behind. They were coming in streams over to us. They were as happy as we were; they didn't want to fight and give their life, their blood. They were jumping in the sand. One big ol' German grabbed me and picked me up, he was so happy. I knew he didn't mean me no harm. He even tried to kiss me. One of my men screamed for him to put me down, but it was all right. I was so glad it was over, because I knew I would live. Another day of fighting could mean a lost arm or leg, or death.

The fighting had stopped in Europe but not in Japan. So they sent three hundred of us Buffalo Soldiers to the South Pacific. The military wanted us to participate in an all-out invasion of Japan. Now I was brokenhearted. I wondered how in the world did they choose me. After a couple of weeks of training we were sent to Japan on a ship. Our job was to take some of the airfields near the mainland. I thought I'd never get back alive. I knew the Japanese were not going to give up easy. I asked the Lord to just let me hit the land—I didn't want to be shot out of the water.

The Lord fixed it otherwise. Three days before I got there they dropped the atomic bomb. We were on a ship on our way there when the Japanese surrendered. I was overjoyed. I knew I would survive this war.

They sent us back to the States and we landed in Norfolk, Virginia. We were all Buffalo Soldiers, and they didn't want us to go through the town. They said it would cause too much traffic, but I knew it was because we were black, and that hurt. We waited there on the docks for two hours with no facilities, and finally our officers said we could walk the five miles to the camp.

As we walked, people were giving us strange looks, as if we were convicts. We were the first troops home, but no one clapped or cheered. The whole town was white and had we been white, they would have mobbed us, they would have been so happy. But things were so segregated, they thought that was how they were supposed to act. Later on I saw how people celebrated elsewhere, but not in Virginia, not for us. The few black people we saw looked scared; maybe they thought they would get lynched or something if they cheered for us. But I was a dedicated man. I was a sergeant and I had one hundred men in my charge. I told them we were not going to walk through town like convicts with our heads down. I said we were going to march with our heads up and shout out in cadence.

We had orders that night not to go into town. Our commander knew if we went, there would be trouble. They wanted to get us out of there. The next day they fixed it so we could go home, wherever that might be. After that we all just scattered. No one ever had any kind of celebration that included us as far as I know, even though we accomplished what all the black leaders wanted us to. When I came home I couldn't even get a job. But while in the Army I vowed that if I lived, I would go back to my father's church, change my ways, and be thankful. That was the main thing in my mind. I had a purpose. I was going to come back and serve the Lord. I wanted to do my share, and that's what I did.

Now when I think about the war, it seems like a dream.

James Hairston,

Navy Steward, 1938–1960,
World War II and Korea

I n the Navy during World War II, no matter how educated or what
skills an African American had, the only rank most could attain
was steward.

Usually a high-ranking officer's personal valet, the white-jacketed
steward would shine his officer's shoes, make sure his laundry was clean
and freshly pressed, tidy up his quarters, and generally be at his beck
and call. Other times, African Americans served as shopkeepers or
cooks, working in kitchens with ovenlike temperatures.

James Hairston was an officer's steward assigned to the USS Hornet,[*]
one of the most famous battleships in history. A survivor of Pearl Harbor,
he was in the middle of many explosive battles in the Pacific during
World War II.

I volunteered for the Navy in 1938 when I was seventeen. I got my
daddy to sign so that I could get in. I'm from Madisonville, Virginia,
and my daddy was a farmer. Back in those days it was the Depres-

*The USS *Hornet* earned nine battle stars for her service in World War II, plus the Presidential Unit Citation. The *Hornet* was in action for sixteen months straight in the Pacific
combat zone.

sion, and the only kind of job was to go into a civil conservation camp.* I decided I might as well go into the Navy. When I joined the Navy I thought it was going to be different than outside, where blacks had to sit in the back of the bus and all, but it was the same thing. It didn't bother me, though—I was used to it.

If anyone my age tells you they were in the Navy and say they were anything more than a steward they are telling you a lie. If you ever asked the Navy to do anything else, they'd want to analyze you. They'd think you were crazy.

I was an officer's steward. I cleaned rooms, shined shoes, took care of laundry, and ran it back and forth, all that kind of stuff. If you had a bad commander, you had a bad time. If you had a good commander, you had a good time. You had to try and make the best of it. As a steward a lot of the guys, no matter what their rank was, gave me a hard time. They thought I worked for them, but I didn't.

I was stationed in Ford Island, Hawaii, at the Naval Air Station. I was over at Ford Island when the Japanese attacked Pearl Harbor. It was early in the morning and I was reading the morning newspaper when I heard a plane going into a dive. I paid it no mind because planes were always flying around the base and I thought it was just a hotshot showing off or someone who went out and got drunk or something. Then I heard another plane and another. A few minutes later I walked outside and saw a yellow bomb hit a hangar. It just doubled up and exploded. I couldn't believe what I was seeing. There were all these Japanese planes flying everywhere.

We had about sixty-five ships, where all the guns were, and they were all burning. All I could do was look. I saw people, hundreds of them, running. I saw people and bodies in the water, and there was nothing we could do. On land we didn't have battle stations or guns. We were completely unprepared, so I just watched. I was shaken. I was never commanded to do anything. All the guns were

---

* FDR's plan to employ thousands of young men during the Depression. They built roads, strung telephone lines, planted trees, and improved millions of acres of federal and state lands and parks.

on the ships, so we couldn't get to them. One plane flew by so low, I could see the bomber holding the tail gun. He was ready to fire right in front of me. Time froze. He looked right at me, but he didn't shoot.

After the attack on Pearl Harbor, I was stationed on the USS *Hornet* in the Pacific, and everyone had a battle station, even the cook. Whenever we got in battle, I had to go and work the bomb magazine. I had to go six stories underneath the water level and load bombs into an elevator that sent them to the hangar deck, and then they got loaded on planes. Every time you went down a level, they closed the hatch behind you and locked it so you couldn't get out. Once you were in there, you couldn't escape. It was this claustrophobic, windowless room filled with bombs. It was below the ship's waterline, so the bombs could keep cool. Once you went down there, you had no idea what was happening above. The only sounds were the bombs hitting the water as they got closer to the ship. The louder the boom, the more you knew things were not going well.

It was crazy down there. When they sent you down there, a chief petty officer went down with you, and he had a pistol on his side. If you got crazy, he had orders to shoot you; you couldn't think. You just had to do your job. There was no time to be scared, and besides, the petty officer and that pistol kept you focused.

Sometimes, men who worked the bomb magazine never got off the ship. If you got hit, the compartment would get flooded and you had to hope someone would come and open the hatch, but the Navy wouldn't sacrifice the lives of the three thousand men above deck to save the lives of the twenty or so guys down there. So a lot of times those guys got left behind. In 1942, the Japanese beat up the USS *Hornet* in the South Pacific. I was lucky: Someone opened the hatches and I was rescued.

I stayed in the Navy and served in Korea, too. Things started to look up for black sailors after that. There were more positions you could have in the Korean War, and I think it was because of the way we served in World War II. We put up with a lot and still did a good job. Now you have black admirals in the Navy, the highest you can go. But if it weren't for my generation, we'd still be doing the same thing.

I stayed in the Navy and I remained a steward. In Korea I was on a ship that carried demolition equipment behind enemy lines. For me, things didn't change much, though—I was doing the same thing. I stayed in because there were lots of good times traveling to different places all over the world, seeing all kinds of things I had never seen before. I also found myself always waiting for the next good captain. Some of the captains were really tough—they were always on you. But when you had a good captain, you could get away with most anything, and they even encouraged you to have fun. Before I knew it, I had put in twenty years.

I left in 1960 after twenty-two years. When I retired I was chief steward. I've been everywhere in the world except Russia and India, and I've traveled to the South Pole six times.

I'm proud of what I did and don't regret any of it. You know, if they let me go back now, I would. After they killed all those people in the World Trade Center, I'd fight again. Yep, I'd do it all again.

Samuel L. Gravely Jr.,

Vice Admiral, Navy Reserves, 1942–1949;
Navy, 1949–1980,
World War II, Korea, and Vietnam

The first job Samuel L. Gravely Jr. had in the segregated Navy of World War II was cleaning sailors' barracks. Despite tremendous obstacles, he became the first African American to graduate from a Navy midshipman's school, the first to serve and command a Navy fighting ship, the first black captain and commander. And in 1971, this son of a postal worker and housewife from Richmond achieved the highest honor of all: Gravely was named the first African-American Navy admiral.

When my mother died, my father didn't want the family to break up, so I went to college close to home at Virginia Union University. Then Pearl Harbor happened and I knew I was going to have to go into the military because there was going to be a draft. I chose the Navy Reserve, because I knew I didn't want to be driving tanks.

After I came out of boot camp I was assigned as a compartment cleaner. I cleaned small rooms with eight bunk huts. Frankly at six foot three, I couldn't stand up; I had to bow down. In any event I was doing that for two months and there was a lieutenant by the name of Stubbs who was the welfare and recreation officer who said, "I don't see many of you black guys playing pool. I want a

black guy to work in the pool hall so that way so many of you won't go into town and get in trouble." I volunteered. I had never been in a pool hall in my life.

Two months later I happened to be in the pool hall, cleaning it as usual, and he said, "Why don't you take the V-12* test." I said, "You don't have any black officers now. Why should I waste my time?" And he said, "Get your butt down there and take the test."

I went down, took the test, and passed. Then I was sent another letter asking me which school would I like to attend. I chuckled because I knew I couldn't go to the University of Richmond, my hometown, or University of Virginia, because they don't take blacks, so I sent it back with no choice. They sent me to USC in Los Angeles and put me in a room with five white guys. I unpacked and was ready to start a new phase in my life. The next morning the administrative officer came up to me and said we are taking you to a school on the other side of town, we can't handle you here, and so I went to UCLA. After that was midshipman's school at Columbia University, and then I became an ensign. I was the only African American among more than a thousand graduates and I got a job training black recruits in marching and simple things like that. Most of the other graduates were assigned to ships to support the war effort.

I finally wound up on the *PC-1264*,† a ship with an all-black crew. Another ship, the USS *Mason*, a destroyer escort, had been commissioned with a completely black crew, too, but no black officers. On the *PC-1264* there were sixty-five black sailors and five white officers, but I relieved one of those white officers. I was the communications officer. I was in charge of electronics, wrote encrypted messages, and made sure that all traffic on the ship got to the correct officer. We did not go overseas. We were homeported in New York City. We patrolled the Long Island Sound and escorted convoys up and down the East Coast from New York to Florida.

*A test designed to screen selected men for college training, followed by a commission as a naval officer.

† A submarine chaser. The ship was commanded by white officers whose job was to train the black crew to handle the ship.

. . .

While stationed on the *PC-1264*, we went to Miami. I had been to Miami two or three times and since I had been stationed there before the black crewmen asked me where they should go for fun. I told them and went with them. Besides, I couldn't socialize with white officers in their bars because of segregation. At about 2:30 P.M. we were all at this bar drinking and the shore patrol walks in. The first thing they do is see me sitting there and they come up to me, knock me off the bar stool and on my butt, and say, "We are going to arrest you for impersonating an officer." They put me in a paddy wagon. My guys were ready to riot, they were so upset. When they got me to the jail they had me walk a chalk line, put my hand out in front of me, and all the other things you would have a drunk do, but I wasn't drunk. They even had me talk to a psychiatrist.

There was nothing I could do. There was no such thing as saying anything to them. They were the authority. They had never seen a black naval officer before. I was held in a jail cell until midnight, which is when my skipper came and got me.

The strange thing was when I got back to the base the next day, the skipper of the base wanted to have me prosecuted and court-martialed for conduct unbecoming an officer because I was in a bar with crewmen. My skipper met with him three times to convince him I didn't do anything wrong. I was in a black bar with crewmen because I couldn't socialize with my fellow officers, who were white. But I've often wondered what happened to those white guys from shore patrol, if they ever got in any trouble for what they did to me. Probably not.

I left active duty in February of 1946 but stayed in the Reserves. I got married and wound up a railway postal clerk. I didn't know if wanted to go back again. I had wanted to take reserve-training cruises but was never offered one; they always told me that all the ships were filled. I didn't think I would come back until the Navy asked me to go on active duty for a year in 1949 to help recruit blacks. Strangely enough, things were slowly changing. As soon as

I came back, the first thing I was asked to do was work at the Washington, D.C., recruiting station. The commanding officer said I could have a room at the officers' quarters, which wouldn't have happened before. Ever since I had been an officer, it was rare that I didn't have to get a room for myself in town wherever I was stationed.

I was happy to be doing something I knew about, even though I had never been a salesman. After a while selling the Navy to high school kids, I had just about run out of what to tell them and decided to go out and see what was going on for myself. I applied for sea duty and got on the USS *Iowa*. The Korean War was going on and they were building up forces. We went to Korea and fired our sixteen-inch guns quite often. We closed up some tunnels where the North Koreans were coming in and also hit their trucks and trains. I felt like a sailor too. I cheered when the guns hit the right spot and cheered when a spotter came back and said you hit your target. I was quiet elated to be a part of it. We were never fired at, because having sixteen-inch guns is like having a battalion around.

All I ever wanted was to get command of a ship, and that was a long road filled with lots of other ship duty and schooling. But finally I got my first command, on the USS *Fulgout*. After that I got my dream: I was next made the commander of a destroyer, which is what I was waiting thirty years for, and it was during the Vietnam War. In Vietnam we would get a call from a pilot or a spotter on shore. He'd give us a target and we'd go "Bang, bang, bang." In two tours alone, we fired eleven thousand rounds. I served three tours.

While I was commander of the USS *Jouett* in 1971, we were between Pearl Harbor and San Diego, and I heard a rumor that I had been selected for admiral—the first black in naval history. I wanted to see it in writing because it was just a rumor and I didn't believe it. I went down to the radio shack to see if anything had come, and there was nothing. Suddenly the guy said, "Wait a minute. Here it comes." I saw my name on the paper as it typed out. I was overjoyed. I just wished my mother had been around to see that. She was the one who pushed me all the way in school and told me to work hard and get a good education, and now I had done as well as she wanted me to.

When I saw my name I cried a little bit. I was working toward just having the chance, just to be considered on my merits. I had problems, but you have to know how to handle it. I didn't have the time to let the problems, the racism, defeat me. All I knew was that I had to be the best I could.

PART TWO
Korea

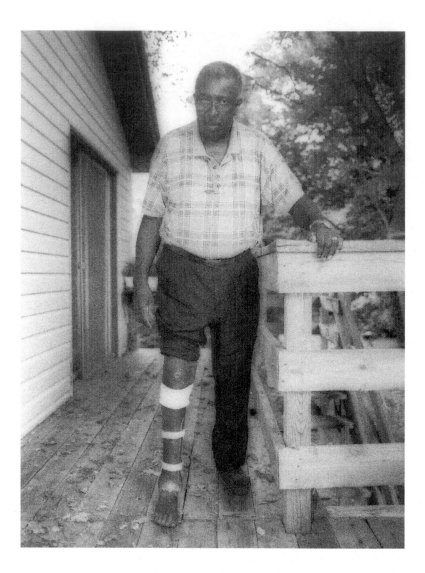

Charles Armstrong,

Army, 1945–1956, Sergeant, World War II;
Lieutenant, Korea

Charles Armstrong left the bloody hills of Korea strapped to a gurney, fighting for his life. After thirteen months of almost daily combat, Armstrong was finally going home. But it wasn't in the way he had hoped. The battle at Heartbreak Ridge, in the middle of the war, had cost him his right foot.

As Armstrong had begun the trek up the mountain where the gory battle was raging, he was shot in his thigh and his femur was severed. He was still conscious after he rolled down the narrow, rocky mountain. The twenty-five-year-old first lieutenant called the medic for help. As bullets flew all around him, he closed his eyes, thinking he was going to die, but when he opened them again, American soldiers were tending to him.

Armstrong, a native of Winston-Salem, North Carolina, saw hundreds of his fellow soldiers in the Second Infantry Division die around him. He was racked by hunger and fatigue, loneliness and fear. He prayed constantly, and believes he survived because of God and sheer luck.

Even now, more than fifty years later, when he closes his eyes, the memories of war come back to him and he becomes once again that skinny, newly commissioned officer of one of the first integrated units, stationed on the front lines in the Korean conflict.

. . .

My job was being a combat officer. I led the troops into combat. I would tell them what to do. The white soldiers had no hang-ups about being with me because they realized they had to follow me if they wanted to get out alive.

Some used the "n" word, but I got over it. One of them said to me, "Nigger, give me a light," and I said, "What did you say?" and he said "A light, nigger." And I said, "I don't play that." Then some of the white guys in my troops told him I was an officer and not to do that to me. That guy became one of my best soldiers. Later on I met that guy in the States and he said my leadership helped him survive.

There were firefights every day. Korea was very hilly, which made it tougher. We had to climb those hills every day on patrol. I would start out with forty men and end the day with twenty. When they were dying they would call my name: "Charlie, Charlie, can you help me?" I would go to their side. Sometimes the weather was so cold it would drop to like thirty-five degrees below zero and their blood would coagulate. They would drown in their own blood. It was a horrible sight. I was scared but I got to a place where I just tried to make the best of it that I could.

The crazy weather was a factor. Winter was so cold, it was like subwinter, and we had to put on layers of clothing, and still it made no difference. I got frostbite and couldn't feel my toes. In the summer the temperatures would go to over one hundred degrees, and if a sweaty soldier took his shirt off or rolled up his pant legs, those big mosquitoes would get him.

We lived like bandits. You didn't get much sleep. Sometimes you stayed up a couple of nights straight; if I got fifteen to twenty minutes of sleep a night, that was good. Sometimes we had rations, other times we had to eat what we could find. One time a guy in my company said a pig was running around and asked if he could kill it. I said you can't but if it accidentally got in your way you could. That night we had barbecue pig. I hadn't eaten for two days and it was tasty, but I got sick with hepatitis and was in the hospital for two weeks.

. . .

One night my captain told me he wanted me to take Bloody Ridge*
at all costs. American soldiers were getting killed there wholesale. I
went over to the ammunition guy to get machine guns and asked
him to pray for me. I had faith and knew God would protect me.

When I told my men that we were going to take the hill that
night, they cried. They told me that we were all dead. They said,
look at what's happening to the other soldiers that are going up
there. I told them that's not us.

We attached bayonets to our rifles, and as we marched, we stabbed
every dead body we walked by to make sure it wasn't the enemy pre-
tending to be dead. Napalm bombs, grenades, and chemical mortars
exploded all around us as we marched up that rugged, barren hill. We
fired in the darkness but we weren't able to see much of anything. We
stayed close together, and I didn't lose a single soldier.

Once we got close to the top, we dug foxholes and settled in. One
of my soldiers volunteered to be on first watch, so I decided to take
a nap. When I woke up it was daylight, and I saw the shiny heads of
land mines reflecting in the sun. Somehow we had wandered into a
minefield without setting one off. It took them six and a half hours
to get us out of there.

When the captain came to relieve me, he said, "I'm putting you
in for a medal." I never got it. Instead, I was sent right back into the
battlefield. This time it was Heartbreak Ridge.† As soon as I got
there I was shot in my thigh, and that was the end of thirteen
straight months of combat. They had to amputate my foot. I was
upset, but what could I do about it? I was lucky they didn't shoot
me in the head after all the fighting I was doing.

---

*The Battle of Bloody Ridge lasted almost two weeks in late August and early September
of 1951. The most intense artillery barrage of the war was unleashed on these slopes. An
unknown GI gave Hill 983 the name Bloody Ridge.

†Heartbreak Ridge is a narrow, rocky mountain mass where thousands of U.S. and
Korean soldiers fought and died in a bloody monthlong battle. The battle for Heartbreak
Ridge began in September of 1951. It was called Heartbreak Ridge because countless
times when the GIs thought they were just about to take it, they were driven back. Their
frustration was heartbreaking—but they eventually took the ridge.

.   .   .

After I recovered, I asked to be assigned to the Army Corps of Engineers but was told they were not integrated yet. I had a degree in architectural engineering and was one of the few black officers to have successfully led a mostly white combat squadron, but it didn't matter to the Army brass.

The Army just didn't know what to do with us back in those days. I was never given a chance to prove myself. I was stationed at Fort Jackson in South Carolina and was assigned office work, flunky jobs. Every day at work was hard, because they weren't treating me right. They gave me all the menial jobs. They treated me like I was nothing.

My wife was having trouble dealing with the stress I was under and one day she burst into tears. I tried to comfort her but didn't have the right words, so I went into the bedroom to change my stump sock and I saw it was soaked in blood. My three-year-old son stood at the doorway watching me. "Daddy hurt, Daddy hurt," he said, and he was crying.

That took me to the breaking point. That moment I knew I couldn't stay in the Army anymore. I just couldn't take it no more. The next day I put in for a discharge. It was granted, and I left in 1956. I went on to have a career as an architect in the private sector, but because of my disabilities I had to retire in 1973.

I still get these phantom pains that feel like spasms when the weather changes. The sound of them bombs and the gunfire robbed me of my hearing.

Sometimes I regret leaving the Army. I was going to stay thirty years, be a role model, become a general, but they found it unbearable to promote me. They passed over me five times. I knew it was a losing battle, but I wanted to serve my country. So few of us were commissioned officers, people were looking to me to be successful. I felt it was my obligation that I fight. I thought I would benefit, but I didn't. They had a way of keeping you in your place.

I want to talk about the Korean War, what I went through, but when I try to tell a story most folks turn their heads, mumble a feeble excuse, and walk away. They say, "I don't mean no harm but I don't want to hear it." Some young people are nasty. They say, "I'm not going to go into the military and get treated like you. You were used."

I still make calls and try to get the Army to recognize what I have done. I'll always keep hoping for a medal or something, but so far it's not happened.

Stephen Hopkins,

Army Corporal, 1950–1953

For four months, through rain and cold and under the cover of night, the American prisoners of the Korean conflict marched from Korea to China. The sounds of gunfire rang in their ears as bullets were fired at those who tried to escape. On the long march, the only way to quench their thirst was to drink from puddles along the roadside. As they passed through villages, old ladies would spit and scream at them. Others would take bamboo sticks and swing at them. Their days were spent in abandoned factories, which were targets of American bombers. By the time they made it to the prisoner of war camp in China, only about a hundred men were left; two hundred had died. Many of the survivors were sick with dysentery and other diseases.

One survivor was Stephen Hopkins, a nineteen-year-old from the projects of Philadelphia. Wasted to ninety pounds, Hopkins spent the next three and a half years living in subhuman conditions. He sustained himself by reciting three Hail Mary's a day and dreaming of his family. Hopkins was one of 7,245 Korean War POWs. More than 2,800 POWs were killed in captivity.

I went into the Army when I was only eighteen. My father died when I was four and my mother raised us during the Depression. I thought it would be a way to bring more money into the house. I joined up a month before the Korean War started and fought for five months before I was caught.

We were on patrol for a couple of days and we had this new captain. He was a gung ho captain who, instead of standing behind us, would lead, so the fellows liked him. This one morning he took the whole company and set us up on a hill and then took a squad of about ten of us to shake up the enemy. There was supposed to be five or six of them taking food from civilians in the village. The ten of us took off across the rice paddy field. When we entered the village we saw what looked like the whole Chinese army. There were about six or seven hundred of them coming out of the village like rodents. We were trapped.

When the company saw the Chinese, they started firing at them, and we were right in the middle. The battle went on for three or four hours. When it was over, we were still stuck in the rice paddy and the Chinese were still there.

One by one, a soldier from the Chinese army would walk up to one of us, point a gun to our head, and say something that sounded like "Move," but all the guys hesitated and were shot in the head. When it was my turn, I just got up and did what he said. Only the captain and I were left, he had been shot in the leg. They separated us. I was with all the blacks. I never saw my captain again. I later heard he was killed in a bombing raid while we were holed up in a factory during the march.

We marched from winter to spring, and it was so cold. I always thought they were going to get tired of carrying us around and shoot us. By the time we made it to the prison camp, we were burying two to three guys a day. Out of three hundred guys, only about a hundred were left. The guys were dying from just about everything. I was always scared. You never knew if you were going to be taken somewhere and be shot. A couple of guys tried to escape, and you saw the Chinese bash their heads in or shoot them.

Our first job at the camp was to catch fifty flies a day. It seemed silly, but it was for your own good. There were all these flies around, attracted to the open sores. The place was infested. We also had to clear roads and set up makeshift playgrounds where we would go for these lectures, which were propaganda speeches. They would have pictures and films of how the black man was treated in the United States. They would have pictures of lynchings and houses burning and things like that. Then they'd ask you why you wanted to fight for these imperialists.

As prisoners, we used to talk a lot about food. I would say that when I got home I was going to buy a whole loaf of bread, peanut butter and jelly, and a quart of milk. Those were the things that kept me going, and I was always dreaming about home. I also got more religious.

I had seen hard days, and I think it helped me to survive. When I was growing up, there was never enough to eat. We had a lot of times when there was nothing to eat at all. Sometimes we just had oatmeal and my mother had nothing. Those things stuck with me. In camp we would get a spoonful of sugar a week and we'd nibble on a little bit of it every day. Some of the guys who'd had easier lives would break down. They would do anything to get a cigarette or an extra bit of sugar.

We would cook in a hibachi pit in the barracks and it was always the same thing—cracked corn and sorghum. If a boll weevil fell in the food, at least you got a meat ration. Otherwise we never got any meat. We slept on the floor. If we put wood in the hibachi it would heat up the floor. There were ten or twelve men in a barrack that was made out of straw and mud, but some held more.

It was joyful when we heard the war was over. They gave us better food so we wouldn't look so bad—we got rice. Fifteen days after the war was over, they let me go. But when I came home I couldn't even eat plain gravy—it would go right through me. I lost half of my pancreas because of two ulcers I had after the war that could be traced to the food I ate in the camp. I had a sleeping disorder that was misdiagnosed as battle fatigue. For thirty-six years, I couldn't stay awake. I couldn't drive ten miles without having to pull over and rest my eyes for twenty minutes. In the middle of conversations I would fall asleep. At night I couldn't sleep well and could only get two to three hours' sleep. The first bit of noise would jar me awake. My disorder wasn't correctly diagnosed and treated until 1986. I've also got real bad arthritis. My feet swell up really bad in the summer. When I went to get benefits and things, Uncle Sam didn't do much for me.

I'm proud of my service. It means a lot to me. Every Memorial Day and Veterans Day I march in the parades, and I'm active with the local chapter of the Korean War Veterans Association. You hear a lot of people saying you don't owe this country anything because of your color and what's going on. But I look at it this way: If you live over there and see how people live, it's a blessing to be here.

Robert Yancey,

Navy Cook, 1943–1947, World War II;
Navy Reserves, 1947–1950;
Army Sergeant, 1950–1971,
Korea and Vietnam

 obert Yancey is a veteran of three wars spanning almost thirty years. He was a teenager when he grabbed a machine gun on the deck of a ship and struck down Japanese planes during World War II. He was a young robust man when he fought in foxholes and laid mines on the front lines as a member of the 24th Infantry in the Korean War, the last segregated Army unit to serve. And he was an experienced military veteran when he served as an Army medic in Vietnam and saved hundreds, maybe thousands of lives.

Yancey enjoys looking back. He's got a shoe box filled with photos and his medals are all neatly labeled and framed. He never misses an opportunity to talk about his experiences, because more than anything he wants folks to know that blacks served their country.

I was drafted into the Navy in 1943, when an African American in the Navy could only be a cook or a steward. I was a cook. I didn't get any basic training because they were hard up for men at that partic-

ular time. I was on a PC submarine chaser and I got my training on the ship. My combat station was down in the hole passing up ammunition. One of the guys said to me, "Hey, Yancey, you don't want to be in that hole when the ship gets hit. They are going to close that hatch down and you'll be gone."

I told the captain I didn't want to be down there. I said I wanted to be on the machine guns. "You don't know nothing about those guns," he said.

"I'm not stupid. I'll learn like everyone else," I replied.

"I'll tell you what I'm going to do," he said. "We are going to have target practice in about two weeks before the invasion of the Philippines, and if you hit them targets you can have that gun."

So lo and behold, in about a week a hyperplane comes in with a 350-foot cable pulling the target. Here's how it works: When the plane gets mid-ship, you start firing at the target, when the target gets mid-ship, you cease firing. Well, I didn't hear the command to stop, and the next thing I knew the plane is trying to get out of my way. So they finally got me to stop firing. "Yancey, you missed the target," the captain said. "But you tried so doggone hard, we are going to give you that gun so you can get the Japanese."

I saw action in the Philippines and I was in Okinawa, where the last invasion of World War II happened. I was shooting at planes. I don't know how many I hit but that's what they call "survival." That's what I try to convey to the youths whom I counsel. I go to the prisons and teach behavior modification, and when I first started teaching in 1976 they said, "Uhh, Mr. Yancey, you are a veteran of three wars. I know you killed somebody. You are no different than we are."

"I beg to differ with you," I said. "A soldier is a warrior, and to be a warrior you kill or be killed. You killed out of passion, and you are going to force me to do the same thing if you don't get out of my face."

After World War II, I was a police officer in Darby Township, Pennsylvania, and I was going to school for criminology. As I was finishing in 1950, the Navy—I was in the Reserves—sent me a notice that said I have twenty-four hours of readiness. I said I'm not going back in the Navy. So I went and enlisted in the Army. They set me to Korea in September 1950. When I got to Korea they put all African Americans in the 24th Infantry.

It was all black enlisted personnel and white officers. In the 24th there were two things basically wrong. Number one, you were spearheading for the rest of the regiment, since you were out front, and we didn't have the top equipment. I was an assistant platoon sergeant with fifty-three men. We'd lay minefields and set up perimeters.

On the front lines you had to dig in and make sure you knew where the enemy was located, that he was in your sights. Before you kicked off an attack, everybody fired machine guns, mortars, everything you had, but these North Koreans and Chinese were like ants— they would dig right through to the other side of the mountain. And nothing was hitting on that side. Then they came in like a plague. Everyone was firing and those jokers were still coming. The Chinese would take five hundred men and go over barbed wire, go through firepower, mines, and booby traps. In the Chinese companies, every tenth man had a weapon. They had to wait for the guy with the weapon to get killed before they could grab it. They had hand grenades and booby traps but no weapons. In an American company everyone had a rifle. We survived because we had the firepower. They had the manpower. Had they had the firepower we had, we would have all been in the ocean. They would have pushed us in.

When you are on the front line, the object is to stay alive; know where you are at and know the location of the enemy. The problems start when they infiltrate the line, when they break the line and get behind you and you are shooting all around. It's crazy.

The more you're on the front line, the more you learn to control your fears. Most people got killed because they were fearful. They couldn't control their fears and did dumb things. You'd say, "Stay down," and they want to look and so they get hit right through the head. Many times after skirmishes, you'd see the trucks loading the bodies in, going to the graveyard, and you ask yourself, When is it going to be my turn? That's the part that was hard. But you learn not to question the man upstairs, you understand me? You don't question him.

We'd go out and lay minefields plus set up perimeters, and I'd tell the guys, "I'm the recorder. I record the minefields. You don't ever, ever go back in a minefield for nothing." One time we were in there taking inventory and so forth, and when we came out I said, "Where is Kincaid?" and then I heard a big explosion. They said he walked back in to get his field jacket and I said, "You see what happens when you don't follow instructions."

I would have got him another field jacket. He lost his life. He weighed about 145, 150 pounds, and they only found about 10 pounds of him. He stepped on a mine. When things like that happen, when part of your crew, part of your squad gets killed, everyone gets shook up. When you are in the military, you don't have your family with you, so you take to the people you are surrounded with. You'd give your life for them and they'd give their life for you.

The conditions in Korea were just horrible and you just wanted to get the hell out of there. It was so cold they called us the Frozen Chosen. They would pull the truck up, someone would let the gas out, and then someone would throw a match and you would see two hundred guys trying to get near that fire. We were the only unit that fought for 126 days straight, all day long. We didn't change clothes or take a bath or anything. We had crabs and lice all over us.

After about 120-something days, they pulled the whole regiment off the front lines and they bought in the medics with large DDT* sprayers. We had to take all our clothes off and throw them in a pile. Then they sprayed us. We went into these trailers with portable showers and then on the other side were new clothes. Then they put gasoline on our old clothes, threw a match, and burned them. That was the end of the 24th Infantry.

*An insecticide.

. . .

I came home for thirty days and then was sent to Fort Bennett in Georgia. Then they sent me to a military police outfit at Fort Bennett, but they still hadn't implemented the order ending segregation, so all the whites stayed three blocks down the road and all the blacks and Puerto Ricans and all stayed at the other end. The company commander was white and the first sergeant was white. When I got down there the major told me, "You ain't got no business with all them stripes, and I'm going to get them."

"Major," I said. "I got them fighting, and that's the way I'm going to lose them."

He told me that I was going to be on patrol in Columbus, Georgia. "I'll give you a corporal as your driver but you won't stop a white soldier for nothing. You are a sergeant in the military police, but that doesn't start until you get in your black district," he said.

I'm thinking, Where in the world am I?

I went to town and two black guys and two black soldiers were mixing it up together, so I shook them down, and as I'm doing that this white officer comes on down, a police sergeant, and he said, "Give me these little boys."

I didn't like his tone but I didn't want to start any problems, so I told the two civilians to go with the officer and I was going to arrest the two soldiers. "I want them all," he said.

I pulled my gun. "You don't get nothing now," I said.

So I got in trouble. My major said he got me now and put me in for a general court-martial for pulling a weapon on a fellow officer. He then restricted me to my room. But I knew he was wrong. I lay around there for thirty days and then I went to see a legal officer. I asked him if I had the authority to apprehend a soldier over the local police. He said yes. I told him the major had me in for court-martial because of what I did and he called the major in and they dropped all charges against me.

So next they gave me a corporal job. They wanted me to go to the stockade and bring in the inmates and show them how to paint pedestrian lines on the street. An infantry colonel came back there, saw me, and said, "Sarge, come with me. You are a disgrace to the NCO* Corps, down there on your knees like a private."

He told me to go into the car and I said I can't; I had to take the fellows back to the stockade. So he came with me and we went to see the major.

He chewed the major out. He told him the Army's changing and if he can't change, he should get out.

They gave me another job as first sergeant, taking care of black guys in my section. We performed the same patrolling duties as whites, but in the black section of town. After a few days the major and first sergeant wanted to see me.

I knocked on the door and no one answered. So I walked in and the first sergeant asked, "Who told you to come in?"

"Let me tell you, boy: You got eight stripes, I got seven," I said. "You get mixed up with me and you'll lose everything, because I'll whip your butt right here. I'm tired of this here. I'm not a private. I respected you and gave you courtesy, and you return it!"

The major said I shouldn't talk to my first sergeant like that. "Sir, I'll talk to you the same way if you disrespect me," I said. "I'm tired of this here. I'm a professional soldier. I'm no private."

They then told me they wanted my section to have a full field display, which is when the officers inspect your clothing, and they wanted my men to bring all the clothing down the road. They didn't want us to have it at our facility, which was the way it was usually done; they wanted us to carry all our stuff several blocks to another facility. It didn't make any sense. I said it don't work that way. They were just trying to make it difficult for us.

I then went straight to the colonel, and he said, "Sarge, we need more people like you to take up for your men." He called the major up and chewed him out.

"Colonel," I said, "I'm tired of this. I've been here for six months and you can send me back in the battlefield."

He said he would send me to Europe. When I got there I was the

---

*Noncommissioned officer.

only black guy there. They had just started to implement desegregation in Germany. They tried to mess with me again there, but I didn't stand for it.

After I spent time in Europe I was sent to a medical training center in Houston, then to Thailand and then to Vietnam. I got reprofiled after the Korean War because I had foot problems from the frostbite, and I became a medic. As a medic I was first sergeant of a medical unit and a hospital administrator. I was in Vietnam for twenty-six months.

I only went out to supervise mass casualties. For example, they had an ammunition dump that blew up and they had sixty-five casualties. They told me not to leave the compound unless I had military escort, but the police didn't come because there was a lot of fighting going on and they were waiting for it to slow down. There were a lot of injuries and I was needed, so we left without any escort.

When I got down there they wanted me to sit in some hut and I said no. I wanted to hit the ground. A half hour later, the hut where they told me to sit blew sky-high.

On the ground, all the wounded were screaming and hollering and I'm shooting them with morphine, putting tourniquets on, and telling the guys where to pick them up. We got the wounded all secured on the truck. They gave me the Bronze Star* because I didn't lose any patients.

You know, that's your greatest reward—to comfort the wounded, take care of the civilians. In Vietnam they didn't have but one doctor for sixty-five thousand civilians, so you can count how many actually saw him. After we had sick calls for our military personnel, then the civilians could come in, sometimes we had two to three hundred a day coming in. When they got an infection they would put maggots in there and the maggots would eat the infections and keep them from getting gangrene. This guy used to come in and we

*Awarded to individuals who, while serving in any capacity with the armed forces in a combat theater, distinguish themselves through heroism, outstanding achievement, or meritorious service not involving aerial flight.

would have to take a syringe and flush the maggots out of his leg so it could heal from the inside out. He would come in every day. We also saw a lot of civilians with snakebites, dog bites, malaria—you name it. I saved many lives and I guess that's why the good man upstairs has saved mine.

I retired in 1971 and went back to college and became a history teacher. I had dropped out of school in the seventh grade. The teacher told my parents to take me out because I couldn't read. Back then I was wearing what they called milk bottle glasses, you know? Now, I even have my master's degree. I taught in the prison and retired as the principal of the prison school.

I went with the governor of New Jersey to Korea in 2002. I felt like I had landed in the wrong place. Before, all you saw were little kids, old people, and dead bodies lying everywhere. Now Seoul is a nice, bustling big city. You think driving in New York is bad—go to Seoul today. We stopped communism. It was an incredible feeling.

I'm a commander of the Joint Veterans Alliance of Burlington County, commander of Chapter 42 of the Disabled American Veterans, and worked on getting the Korean War Memorial in Burlington County. It keeps me young. It keeps me active, mentally and physically. It's our responsibility to carry on the legacy. If we don't carry it on, who will?

Julius W. Becton Jr.,

Army Lieutenant General, 1944–1983;
Army Reserves, 1943, 1946–1948,
World War II, Korea, and Vietnam

t nineteen, Julius Becton was a second lieutenant in the Army. It was 1945, the tail end of World War II and the beginning of his illustrious forty-year Army career. Becton served in three wars, during which he fought and won the battle against discrimination.

Becton, of Springfield, Virginia, was the sixth African American general in United States history. He was driven by his belief that black Americans in the military could thrive and make the country a better place for all Americans.

When President Truman signed Executive Order 9981 in 1948, which ended segregation in the military, I was in the Reserves and in training at Aberdeen Proving Ground in Maryland. I remember the post commander assembled all the officers and he read the order to the assembled group. He then said, "As long as I am commander here, there will be no change."

I wasn't surprised or upset. It was typical Army leadership—Jim Crow and segregation were here to stay. That was the sentiment of the Army at the time. Blacks could not lead. There was not much of anything they would let us do.

At that time I was in college studying to be a doctor, but I really didn't want to be one. That was my father's dream. He only had a third-grade education and my mother dropped out in the tenth grade. I was married and expecting a youngster, and my stipend at school was not enough to meet my family obligations or a mortgage. So in 1949 I went back on active duty and was assigned to Fort Bliss, Texas. I was the mess officer.

One time our unit drove from Texas to Fort Bragg in North Carolina. The first night out in Van Horn, Texas, I had to write up a list of supplies. The men had wanted certain things we didn't have. So some of the other officers and I went into a drugstore and started buying what we needed. We spent hundreds of dollars, and the store manager was happy. So I sat down at the counter to eat an ice cream cone and the sheriff, who was in the store, turned pale and said, "Boy, get away from that counter." I was like, Who said that? I turned and saw a great big badge and a .45 on a scrawny little fellow. He then said to my major, "You need to teach that boy something." I paid for the ice cream and left. My major told him that I wasn't from around there and didn't know. But I just wanted to get back to my unit as soon as possible. There was a certain comfort in numbers.

I went to Korea in 1950 with the 2nd Division. It was all white except for two battalions. The black units were pulled out and not deployed with the regiment. We were sent somewhere else. They weren't sure how we would do. They sent us to Pohang-Dong on the east coast and we saw some minor action. Eventually, they sent us back to the regiment because they needed personnel.

Korea was very cold and mountainous. There was no place to go to get warm. You would stomp your feet and clap your hands to try and get them warm. We were not prepared for a winter like that, because we didn't think we would be there that long.

One time we were fighting on the crest of Hill 201. We couldn't go forward or backwards. We were hunkered down and held our own. You have fear, but you have to control it. I don't want anyone around me who does not have any fear, because they do stupid things. We could not even see the enemy—all we could see was the gunfire.

I got hit with shrapnel from mortar rounds in the right thigh. There were about eighteen to twenty of us and many of us got hit. I was sent to a hospital in Japan and stayed as little time as I could because there was a rumor we would be home for Christmas. So I rushed and rejoined the regiment at the 38th parallel in Korea, just south of Kunari, about thirty miles south of the Yalu River. I was leading an element up a hill and got shot in the Achilles' heel between the tendon and bone. I was very lucky. That wound saved my life, because thirty-six hours later my battalion lost all of its officers except a handful. They were either killed or captured. I was sent back to the hospital in Japan and the nurses said, "Not you again." This time I didn't rush back. It was clear we were not going home for Christmas. The Chinese had entered the war.

I returned in February of 1951. At that time the unit was integrated. No longer was there an effort to keep blacks together. I was the only officer left from the original group and became the company commander. I held that position until I rotated home to Camp Edwards, Massachusetts.

Camp Edwards was integrated, but I joined an all-black service unit. Our job was cleaning the post and running the major mess hall. We had about five hundred soldiers and we ran the kitchen, ordered supplies, and followed the master Army menus. None of the other people at the base had been in Korea, and here I was with two Purple Hearts* and a Silver Star† doing cleaning work. I

*The Purple Heart is awarded to those who are wounded or killed in action.

†The Silver Star is awarded to a person who, while serving in any capacity with the U.S. Army, is cited for gallantry in action against an enemy of the United States while engaged in military operations involving conflict with an opposing foreign force, or while serving with friendly foreign forces engaged in armed conflict against an opposing armed force in which the United States is not a belligerent party.

didn't feel good about that. I had just spent the better part of a year in Korea, in which I was fighting for my county and fighting to prove I was qualified. The president had signed an order basically saying the military was desegregated, and I had come to an all-black mess unit. I knew there was a better way some place, and I was going to find it. I put in for a transfer and got out. I got reassigned to Indiantown Gap Military Reservation and was involved in basic training.

My career kept going along. I was promoted and took every opportunity to learn.

During the Vietnam War, I was the commander of the Airborne Cavalry Squadron. Our job was to be the eyes and ears, or I would say, "Sneak, peek, and be damn cautious." I got another Silver Star and a couple of Bronze Stars.*

In Vietnam there were not a lot of senior black officers, and that created the wrong impression. About 20 percent of the enlisted men were black, yet less then 10 percent of them were officers. It created the impression that the Army didn't care. People like to look up and see someone they can relate to. You need to have blacks, Hispanics, Native Americans in proportion to what is in the force, which is why so many of us believe affirmative action is so important. We do not need to have units where there are no black officers or just one with 20 percent black enlisted men.

When I left Vietnam I was selected to be colonel and sent to the National War College. When I graduated in 1971, there were no black generals. In 1972, I was one of five blacks selected to be brigadier general. I was the first in that group to be promoted and the sixth general in Army history.

---

*The Bronze Star is awarded to individuals who, while serving in any capacity with the Armed Forces of the United States in a combat theater, distinguish themselves through heroism, outstanding achievement, or meritorious service not involving aerial flight.

I retired in 1983. It was a mandatory retirement. I didn't think I was ready then but I do now. I'm very proud of what I accomplished. We were pioneers. We saw the good and the bad, we saw the Army go from its segregated past to where it is today. Now I can't think of a finer place for black Americans to be.

PART THREE

Vietnam

Norman Smith,

Marine Sergeant, 1964–1967

An eighteen-year-old high school dropout from Philadelphia, Norman Smith was so impressed by a TV documentary about the Marines that he immediately signed up. But as soon as he got to boot camp, he was sent to Vietnam, a country he had never even heard of. He served two tours and is among one of the ninety percent of Vietnam veterans who say they are proud to have served their country. Now a hardware store employee, Smith says his war experience made him a better person.

I think about Vietnam quite a bit. That kind of experience stays with you. I'm patriotic, which comes from being a Marine. I think this is the greatest country on earth. The Marine Corps sent me to such shitholes of the world that I have gained quite a bit of respect and admiration for this country. In Vietnam they had open ditches for sewers and ladies with thick pants would squat, pull up one leg of their pants, and urinate, right out in the middle of the street, in the gutter. I'm sure they live better than that now, with all the money we threw over there.

·   ·   ·

I served in Da Nang and lost a couple of friends in my first tour. One was a cat, Jordan, from New York—he was a brother—and the other was a white boy named Bugs.

We were on patrol as a company,* we're talking about seventy men. We stopped for a break and took our helmets off. Afterward we stood up and were about to cross a river near a rice paddy field and continue our patrol when . . . well, some kind of mine went off.

I was further back and the word traveled back that some guys were down. Then the names traveled back—Jordan and Bugs were hit. I rushed to where they were down and when I got there Jordan was dead and they are working on Bugs, but he eventually died. Bugs had a big gash on his head; a piece of shrapnel went into his forehead. Jordan had shrapnel wounds on his upper torso; it looked like rice and I really don't know whether it was rice or shriveled flesh, but it looked like rice. Those was our platoon's first casualties.

I cleared Jordan's rifle, took all his bullets, and kept his hat. I had his stuff in a plastic bag, with his funeral card, and that was going to be my keepsake. I would still have it, had I not lost my bag on my way to Vietnam the second time.

On our first patrol, after Jordan and Bugs were killed, we were sweeping though a rice paddy in this village and we took rifle fire. So everybody got down and we tried to figure out where the rifle fire was coming from. These guys were in a hole that went down, back under the hill, and into a horseshoe. On top of the hole was a wooden frame and a plot of dirt that sat right down on it. You could pass it a hundred times in a day and never know it was there, but some dizzy stuck his head up out of the ground while we were sweeping the area; he thought we would pass but instead he looked straight into one of our faces. Then he dropped back down in the hole, and now we knew where they were.

First thing we did was told our interpreter to tell them to come

*A unit consisting usually of a headquarters and two or more platoons.

out. They didn't come out. Somebody popped a smoke grenade in the hole. A smoke grenade will burn fifteen or twenty minutes. We didn't know the configuration of the hole they were in. We threw the smoke grenade in because if there was a back entrance to this tunnel, we would see the smoke come out and know which was the back and front entrance. So we don't see any yellow smoke coming out anywhere, but we hear these guys in the hole gagging and choking. So again they call the interpreter over to try and talk these guys out of the hole, and they still are not coming out, and then one of them tried to throw a grenade out, but it fell back in the hole. The blast knocked the commanding officer down.

Next thing, a guy stuck his head out of the hole, gasping for breath, and then he dies. Now we still hear someone choking; the interpreter again tries to tell them to come out, but they are not coming out. The lieutenant stuck his arm down there in the hole with a .45 in his hand, big bullets, and he unloads two magazines, ten to fourteen rounds. But there are still some of them back there gasping for breath and they still wouldn't come out, so we blew the hole and that killed everybody.

Then we got down and started digging. So I'm in the hole digging and the shovel hits something. I wonder what it is. I brush the dirt away and then I realize it's a dead body and we pull him up, but by the time they got him up he was naked. They were all naked—the dirt pulled off their shorts, which was all they were wearing. There were about six men in the hole.

It was the first time we were out since Jordan and Bugs were killed, and we were after blood now. We burned the village, took one of their water buffaloes, and tied it to a stake in the middle of the courtyard. It had its little one there and we took potshots at the buffalo until it died. There were families—very few men, no one our age, just old men, young children, and women. All the guys of fighting age were gone or maybe they were in another hole that we didn't uncover. If that guy hadn't stuck his head up, he might have killed some of us. If he gave us five more minutes, we would have crossed their position and our backs would have been to them.

We were angry. We deprived them of whatever they had. We took their rice and everything we thought was of value to them. We took and destroyed it. These guys were the only ones we killed. There was no massacre. These guys were combatants, and they paid the ultimate price for it.

We isolated the villagers. They were taken to one area and someone watched them. We went to each hooch looking for weapons, seeing what we could find, and then we destroyed the place. I can't say I'm proud of it, but we didn't initiate it, they did, so we felt justified, I guess.

After I came home the first time, I drove across country with two white Marines and stopped in New Mexico for breakfast. We went to a place where the three of us hadn't been before, and the waitress came up to us and asked to speak to one of the other guys. They have this conversation and the next thing, we're leaving. The waitress told my friend that she didn't want to serve me. Being the Marines that they are, after all we had gone through, they weren't going to eat there. I was shocked—I hadn't been back in the country twenty-four hours.

When I came home, I got married, took my wife to Parris Island* and then I got orders to go back to Vietnam: The Marine Corps is getting their ass kicked over there now and I'm experienced in infantry and they need ground troops.

If you are an E-4 corporal or above and are reassigned overseas, but have less than a year in your enlistment and have no intention of reenlisting, you can ask to be reassigned stateside and not go overseas. I had less than a year to do, but Washington wasn't even listening to that. I went home and I saw my congressman from north Philadelphia, who told me to go to Washington, but I didn't know anybody there. So I go see another congressman and he said he'd do what he could. But nothing happened. I can remember

---

* Marine base in South Carolina.

sitting in Philadelphia Airport waiting for the plane to pull out, and I cried. I didn't know if I would make it back. I just didn't think so. There was no question about not going. I wasn't going to go to Canada. I was Marine, a good Marine, a dedicated Marine, so even when they sent me back the second time, there was never a thought in my mind about not going

I went back to Da Nang, which is where I was before. I was stationed in a security platoon that protected where the officers lived. It was a stroke of luck: The whole country was dangerous, but I wasn't in actual fighting every day.

Then the word came down that anyone on a second involuntary tour couldn't go back to the States, but they could ask to be reassigned anywhere in the Pacific. I had a good time when I was in the Philippines, so I wanted to go back to the Philippines, and I put in new orders. But when the orders came back, I was told to go out to the front in Vietnam, where the Marines were getting their ass kicked every day. Here I am waiting to leave the country and they are sending me to the 5th Marines on the front line. My bubble busted.

So I jumped on the helicopter and they sent me out there. A regular Marine squad is fourteen men and they had squads out there that were down to five. They were searching the rear areas for anyone with infantry experience.

I get there and check in with the commanding officer. He looks at my records and says, "You are on your second involuntary tour." I say, "Yes, sir. I was waiting for orders to leave the country, but when the orders came, they sent me out here."

And he said, "Oh, I can't use you."

I say, "Can I go back?"

He said, "Do you want to wait for the helicopter or do you want to walk?"

I said, "Okay, I'll wait for the helicopter."

But then he said they only get one helicopter in a day. So I had to spend the night here, and don't you know they got attacked that night. I had to jump out of the barrack, put my flak jacket on, and

I'm praying in a foxhole waiting for this helicopter to get me out of here. I survived that one. Shit was dropping all over the place. Oh, man, they were shooting mortars in. All you could do was get in the corner of a foxhole.

The helicopter came the next morning and I went back to where I was before. I then got orders to go back to Okinawa, which is an island off the coast of Japan and a regular base.

If it weren't for the war, I would have made a career out of the Marine Corps. I was dedicated and I liked it, but the war scared me out. When it was time to go, I got out. I lived through war, and it's an experience that you never forget. I'm a better person for it because I've been there. I value my life and everyone else's more. After you live through an experience like that, you don't want to deal with anything petty. If I don't like you, it's not because of your skin color, it's because you have done something to me personally. I'm not for pettiness. I see the world differently. These little so-and-so's running round the street and shooting each other don't really understand what life is about. I'm not sorry that I went—it added something to my character. But you can't tell that to the mothers of the fifty-eight thousand who were lost.

Elizabeth Allen,

Army Nurse Corps, Captain, 1967–1969;
Army Reserves Major, 1969–1982

Over five thousand Army nurses served in the Vietnam War. The average age was twenty-three and they were mostly all new to nursing. Elizabeth Allen was an exception. She had her master's degree when she signed up and wanted to get her Ph.D. in the Army.

Allen served in Vietnam during one of the most deadly periods of the war—the Tet Offensive. For several thousand years the Vietnamese Lunar New Year has been a traditional celebration that brings the Vietnamese people a sense of happiness, hope, and, peace, but on the Vietnamese New Year in 1968 the North Vietnamese Communists launched a hellish attack on the South. Thousands were killed. Although the Communist victories were short-lived, they fueled anti-war sentiment in the United States because the bloody takeover was covered extensively in the media.

In 1968, the bloodiest year of the war, Allen and the other nurses stationed in the jungles of Vietnam tried to save young men whose bodies were blown apart in the middle of a war zone. Allen, of Ann Arbor, Michigan, is now an associate professor of nursing at the University of Michigan and a mentor for troubled youths. She never wanted to leave Vietnam.

· · ·

I had to go to Vietnam. I needed to go. I made a quick decision on my own. No one knew about it. There are extremely few minority folks in health care. There are a lot of aides, but in terms of professional folks, very few. At that time I had a master's. I also knew that African Americans were most likely to end up in battle units, in the death units, and I really wanted to do something. It had nothing to do with whether I agreed with the war or not. Folks talk about the love of God and country, but that never factored into my decision. What did factor in was that there was someone in need. My three brothers were also in the Navy at the time and I truly believe that every person who claims to be American has military obligations. I don't believe women should be exempt.

I left home at the end of January 1967 and did six weeks in Fort Sam Houston in San Antonio. I wanted to go to Vietnam. I didn't want to sit around. Soldiers were dying in Vietnam, not in America. They kept saying I could go but then they'd try and talk me out of it. They had few people in the military with advanced degrees and they had a nursing program with the University of Maryland and they wanted me to teach there. I had to fight to go to Vietnam, and finally they sent me to Cu Chi Province, to the hospital at a base camp with the 25th Infantry Division.

The Vietcong had been at war since the 1940s and they had a massive tunnel system. Our camp was built on top of them without the Army knowing it. So at night the Vietcong would come out of the tunnels and people would get killed. At first they thought the Vietcong were coming from the outside, and then someone discovered their secret. When I got there they knew about the tunnels. I get there on a Friday afternoon and we were getting orientation on Monday, but we got hit on Saturday. We got hit a lot. If you were off duty, you went to the bunker. If you were on duty, you kept working.

At Cu Chi, people would get caught up in the big wounds, but there were massive psychiatric wounds that I was seeing every day. The Vietnamese people were incredible warriors. They were smart as hell and they used what they had. When we discovered those tunnels, that meant someone had to go in there and get those people out. The Vietcong were half the size of the U.S. soldiers, and those passageways were small and dark. Keep in mind that they opened out into the jungle of Vietnam and they were full of bugs, spiders, and snakes. The Vietcong would hang snakes from the tops of them and so when our guys would shimmy through the openings, the snakes would bite them. But as bad as that was, the worst was when our men would get out of them in the middle of the jungle and these huge spiders had built webs over the mouth of the openings. It was dark and they had no lights. I had more than one soldier whose face hit the belly of one of those huge spiders. They would lose it. Sometimes they would walk into the tunnels and fall into these pits filled with pointy sticks that would pierce them all over. By the time we would get to them, they would be infected all over.

There is another part of what the Army nurses had to do called "medevac." It's done in a helicopter, and that means you have to fly with the troops who are in danger of dying. We moved troops from the field to Saigon. When I got to Cu Chi, I was a captain with no military experience. But in no time it was my turn to fly medevac. I didn't know what to do. They told me when you get the call, you have to go to the helicopter pad. I said to myself, Shit. I haven't been here a month and my number got picked. They told me to sleep in my clothes, be able to wake fast, and go wait on the pad. So at some ungodly time in the morning like one A.M., the call comes. So I had to wear my flak jacket and double pot [helmet] and stand in the dark. The helicopter comes with no lights and the pilot says, "Come on." and I jump on. The minute I get in I ask him, "What do I do?" and he says, "Don't you hear him?"

I then hear this awful sound. It was a young man with a sucking chest wound. The injury has penetrated his chest cavity and air was coming in from the outside, compressing his lungs. All I could hear was this noise. I couldn't see him but the pilot said we had to fly with no lights on. So all the way from Cu Chi to Saigon I've got to keep him alive in the dark, and the Vietcong are shooting at us. It's just the two pilots, the gunner, a dying kid, and me. You think that's tough? We finally set down, get him off, and then take off back into the war zone.The choppers were a big part of Vietnam. If I wasn't flying medevac I would meet the choppers when they came in with the wounded and get them triaged. The hard part is guys who don't get to go to surgery. The more severe the injuries the less likely he is to go to surgery. It's hard to look at those kids and say you don't get to go to surgery. We only had three operating rooms. We took first those who would use the least resources. We didn't have that many surgeons, beds, antibiotics, and blood. So whoever needed twelve hours of surgery was not getting in so fast. Some died and some we just maintained until they got a chance.

After a few months in Cu Chi, I was sent to Pleiku, the first hospital to get bombed in the Tet Offensive in February of 1968. A lot people got killed in the hospital. So they worked experienced nurses at night, and we got hit every night. There was no place to go. You had to keep doing whatever you were doing. Responding to fear is not always an option. Men's lives were dependent on me, and my being scared was not useful. You had these guys with massive wounds—not just a leg cut off, but as bad as two legs off, two arms off and blind. He can't move, and my being scared is not helping him. I had to protect him. I had to make sure he didn't bleed or choke to death. I had to make sure if a mortar hit, the shrapnel didn't hit him again. During the shellings, whoever is the head nurse is the HMFICC—the Head Motherfucker in Complete Charge—and you got to ride it. I had to keep them alive to the best of my ability. I sat on the razor's edge.

I knew the Tet Offensive was going to happen. There were terrible disadvantages of being a black woman in the war, but advantages, too. The black troops seek you. The troops had intercepted communications about the Tet Offensive and they told me it was coming. The first round came in, a rocket at two A.M., and I heard it from a distance. It was a horrible sound and I said, "Good God, we're being hit." And as I said it, the hospital got hit and then there was a steady barrage for an hour. At the time there were only two nurses that were combat-trained and I was in surgical intensive care. We used radiophones then, and I got called. The director of nursing was screaming, "Captain Allen, Captain Allen. You have to go to work." I said, "Will someone walk me to my unit? It's like three A.M. and the rockets are still coming in." After a silence, she said, "We don't have anybody. You have to walk by yourself. Don't forget your flak jacket, steel pot,* and be careful." When I hung up I thought, If I could walk across that field I could do anything.

When I came out, the hospital was on fire, and then I saw the unit for surgical intensive care. I opened the door and this poor nurse, she was about twenty-two and she was so frightened. Some of the guys had fallen out of the bed and wounds had broken up, stomachs were open. They were bleeding. It was bedlam. I walked in there knowing I was in charge and I had to make that work, and I did. I was good.

For the next two months we were attacked a lot. I had nurses sleeping under the GIs' beds. When we got hit we would get the wounded on the floor, get them out of the line of fire. They'd have casts on their arms and legs. They'd have IVs in the neck. They'd be blind with drainage tubes and a catheter and you had to get them out of their beds and onto cots on a cement floor.

The minute they hit the cold cement they had to pee and that meant someone had to help them with that. I made sure they had blankets on the floor, which lessened the chance of them peeing. Guys who were really badly injured, I would put them down on the floor when I came on. They weren't whiners. They could be hurting and bleeding and let me tell you something, they didn't whine. One time we could hear the Vietcong, hear the guns clicking; they were that close. I was under the bed with a kid who was blind and couldn't walk. He was a kid, about nineteen. He was patting me on the leg, telling me, "Don't worry, Captain, I'll protect you." I said, "Don't you worry, baby. That's all right."

*Military helmet.

That was every day, every day—shit. War is war. It ain't what these people think it is. We get all caught up about 9/11, it's a terrible thing, but at least it was quick and we didn't have to face it every damn day.

I stayed in Vietnam for one year. I didn't want to come back home. I wanted to work with the troops. I could be a nurse back in the U.S. any day. I was good at what I did, and the black troops needed me and I needed them. Eventually I did go back to Valley Forge and taught at the Army specialty school at Valley Forge until January of 1969. I wanted to go back overseas but the Army didn't want me to. I was up for major. I wanted to be a nurse consultant and was told those jobs were given to those in the military for a long time. I wanted to be regular Army but it seemed like it was not going to work for me. I was offered a consultant job in civilian life and I got out. I love the military and I would recommend it to anyone, but it wasn't ready for me at the time.

War is not like picking up a carload of casualties on the highway. There's nobody to call. It's you and your skills and your heart— that's it. When the IVs run out, damn, you hang whatever you've got, because you've got to keep the soldier alive until you can get him out of there. We had all kinds of stuff like malaria, kids with 107-degree temperatures and with nothing to cool them down, massive body burns, chemical burns that take their skins off and their bodies are just weeping plasma on the floor.

We talk about war as though it's a trip to the fucking supermarket; well, it ain't no walk to a supermarket. You hear them talking about "We" going to fight—well, who is "We"?

I was on a talk show once and they said, "I don't understand why you all don't get over it already." It's just not that easy. Vets shouldn't have to eat out of trash cans or worry about health care. It's not right, and it bothers me every day.

James Robbins,

Army Lieutenant, 1966–1969

n the jungles of Vietnam, it wasn't just the foot soldiers who were
in constant danger. Support units were also in the midst of the horror of war. These soldiers served in engineering battalions, building bridges and roads. They worked in maintenance and transportation.
James Robbins of Norman, Oklahoma, was one of thousands who served
in radio battalions. These men provided communications for the troops
through thousands of miles of jungle, mountain ranges, and coastal low-
lands that were crawling with the enemy. Radio battalions were constant
targets of the North Vietnamese, because the communications lines linked
American troops throughout the country.

Although he survived the war, Robbins said he died in Vietnam. He
never reclaimed his life.

I finished college and was working at Liberty National Bank in
Oklahoma. I was hired as an executive in training, with projections
of being an officer, which would have made me the first black bank
officer in Oklahoma. I went to work realizing my draft number was
coming up in the not-too-distant future. I decided to enlist. My
intent was to enlist in the Air Force, but the Air Force recruiter was

out to lunch. Not being particular, the Army recruiter was there, so I enlisted in the Army.

Initially, I was to be a radio communications officer in Vietnam at Long Line Battalion South. There were two bases that were responsible for communications links throughout the country and the rest of the world; one was in the South and one was in the North. I was first sent to the South, which was on a rocky mountain, and that was where I was introduced to Agent Orange. I just recently found out that I have Type 2 diabetes, which is recognized by the Veterans Administration as a result of being exposed to Agent Orange.

You see, we were up on this mountain and the area had to be cleared of the jungle because that's where radio transmitters and antennas that transmitted to other locations all around us were. But the jungle would encroach on our compound from time to time. These monkeys would hide in the vegetation and they would throw things at us; the soldiers would fire at them. If you heard them there you didn't know if it was the Vietcong. The jungle grew so quickly, they sprayed Agent Orange to kill it. One minute vegetation would be growing close and the next minute it would be dead. In those days people were saying Agent Orange was perfectly safe, but now we know that's not true, and so many of us were exposed to it on a regular basis.

After about three months in the South, I was told I was being reassigned as company commander to Company C at Long Lines North. When I got there I was picked up and a driver took me to my unit and we parked my jeep. I was a first lieutenant by then and I was being sent to fill a position of a major without a promotion. They were short of officers. The driver then took me and showed me where my quarters were. I went in, took a shower, and went to sleep. While I was sleeping, we came under attack. We were thirty miles from the Cambodian border and I didn't know it was one of the deadliest places in the world.

Someone came to get me and took me to a bunker, which was

buried in the ground. There were thirteen bunkers that held communications lines, a hospital, an artillery unit, and I was responsible for them. We were under attack and took a lot of mortar rounds and I was in a daze; I was shocked. I hadn't had that experience. When we were able to come out, my jeep had taken a direct hit and existed no more. I think at that moment, when I saw my jeep, I resigned myself that I was not going to survive.

Still, my job and duty was to command. I had about 160 men under my command and a first sergeant assisting me as well as six or seven noncommissioned officers. I can only remember the name of two of them. I have blocked out so much. In previous wars, when a unit served, you would go to a war zone as part of a company or platoon or brigade and you stayed with them and then everyone came home at the same time. As a result, the people in that experience had a commonality and were comrades. In Vietnam we served in rotations, meaning each person had an individual time to come and go. There was always people coming and going. When I got there, people I knew who were there with me for three months would leave. I never had the opportunity to develop a sense of connection with anyone. And then if you did, you had emotional trauma when they rotated out and were gone. It was such a fragmented process, which meant you always had new people, people who were totally inexperienced who were expected to perform along veterans, and you had distrust and fear of these new people.

I was twenty-five, the old man. I'd already graduated from college. Most were eighteen to twenty-three.

The bunker was a command center. It was a steel box that was buried in the ground. It had communications lines so that I could have radio contact. It also had water and medical supplies.

I was never told to kill myself if I got caught, but it was made clear how much damage would be caused if the codes of information I had were obtained. Being a gung ho military officer, that was the choice I would have made if given any choice. I was only issued a .45. I didn't have a combat weapon, just a pistol. I wasn't in a position to do much defending of myself, but I made sure I had an M-16.

One time we came under attack and there was this new guy, Dennis, who had just got there and didn't even have a weapon yet. He didn't know what to do and he followed the rest of the guys to a bunker. He was terrified; he wanted to get out and run. One of the officers communicated to me that the new guy was freaking out. I was under orders not to leave the command post because I had codes and was not allowed to risk being taken by the enemy. But I felt such a sense of compassion for this young man that I decided to leave the bunker and go to where he was.

I crawled there, and he was fighting all the guys. He turned to hit me and then he recognized me as the company commander. I comforted him and calmed him down, and from that day on, whenever he was not on duty, he guarded my quarters. We became very close friends. He was the only friend I had in Vietnam.

We called our base the "Aiming State of Vietnam" because it had eight to ten antennas that were 60 feet wide, 120 feet tall, and had little red lights on top so planes wouldn't hit them. The Vietcong used them as a way to zero in on our unit and the units around us.

They were constantly trying to take our site out. It was one of the most hazardous places to serve. We were constantly under attack. Once we were under attack nineteen days in a row with mortar shells and attempts to overtake the compound. We had a barbed wire fence, machine guns, and mines to stop them.

On one occasion when they tried to overrun us we had to do "level fire"—we had to fire right above ground level. The next morning, when we took stock of what happened, we found five Vietcong caught in the barbed wire. One was a man who'd served as a barber in our compound.

In that day and time there was a lot of talk about officers not getting killed from enemy firing but from fire from their own compound. I know when you tell people to do things that put them at risk, bad things can happen. But I made sure my men's needs were met. We had problems with drugs, marijuana, and there was this brown liquid they put on their cigarettes that made them crazy.

The military provided me with a liquor allowance of seven bottles a month. You would get a card, and I had saved them up since being a commissioned officer. So I had one hundred bottles and I used my cards and set up a little club. I chose a noncommissioned officer to run the bar and we stocked it up for the guys to have a place to drink. It was quickly evident that liquor was more predictable than an officer who is high on drugs. With drugs, people could seem fine, then in combat their ability could be greatly impaired by the drugs. I specifically set out to encourage my men to drink instead of taking drugs. It was good and bad in some ways.

One night one of my troops got drunk as a skunk. He was so darn drunk he didn't know what he was doing and I told him to go into my hooch, because he couldn't make it to his own and we were under attack. When it was over he had puked all over the place. We had to have my room fumigated. He had to spend a lot of time cleaning it.

Our latrine was fifty-five-gallon barrels cut in half. To clean them out we would pour diesel fuel and burn the waste, then take a hoe and rake it back and forth, then put more fuel on it until it burned to a crisp—and that smoke, it would get in your hair. We also had big rats, and they would get in there and then you'd run the risk of them biting you. When you got close to the latrine you had to stamp on the floor to chase them. It was awful. Well, taking care of the latrine became his job. It wasn't good to mess with the company commander.

I was always a loving and caring person, but that died in Vietnam. My heart froze. You stop letting your heart be available. I can't remember the names of the men I lost in battle, but I remember writing the letters to their families. I remember one of my men being hit and all that was left was his right arm and when I got to it his hand was closing.

. . .

Being out of control of one's life is devastating. I detached in Vietnam in order to survive. To survive there I became robotic, and that robotic behavior has followed me through life.

Some years after Vietnam I wrote an essay to myself as to why it's been difficult to really belong to this environment again. I think it's because when you are aware of your imminent demise and give yourself over to it, and know you are going to die, life will never have the same meaning. I taped recordings of myself so that my parents could hear what was going on and they would know how I died. My mother kept them, but I haven't listened to them.

We didn't have counselors to help us make reentry. I left Vietnam on Friday evening and was home Sunday. A week later I was back at my job. No one talked to me about it. It was not a popular war and we returned to a hostile environment. A lot of people opposed it, so best thing was to not let it be known that I was a part of it. I said to myself that I was getting on with my life and don't think about it again. But it's influenced everything I've done since then.

I have had so few people that I can identify with and share this with. I never met another African American officer who served in Vietnam. I've gone to veterans groups, but there is a mistrust of officers.

Being older, getting away from the experience, I'm filled with resignation for my life. I wanted a doctorate but I will not get one; it's not feasible for me to put the time and effort in. I had hoped to be a department or division head someday, but I am in mid-level management. I lost three years of my life. I could have been a principal in a school or I would have gotten my degrees earlier in life and got a chance for other opportunities. I had the mentality and the intellect.

When I came back from Vietnam, I did not have the drive. It was hard for me to exert the kind of energy necessary to achieve the things I wanted. Now at the end of my career, I don't want to be angry and envious of people who achieved more, but I am. I'm in my second marriage. My life has been tremendously complicated by suppressing so many of the emotions I had in Vietnam. There are times when I have not been available to people, affectionately connected to people, because of my detachment, and in other ways I have gone overboard trying to be normal. When I came back from Vietnam and realized I wasn't going to die, I wanted a significant other and to a have family. But then I heard I was exposed to Agent Orange and maybe couldn't have children, so I married a woman with children. But if I had not had the Vietnam experience I would not have made that choice.

The marriage didn't survive. I did have a son and have a connection to the two girls I raised, but we are divorced and now I'm married to my second wife, who has four children. She is German, and two of her children married African Americans.

We raised one of her grandchildren, Justin, who is of mixed race, since he was eleven. He is so close to me; we have a very nurturing and loving relationship. He's my boy. I live with a lot of tension but he calls me every day. He's with the Coast Guard. I'm real proud of the man he is. I needed an intense relationship between a father and child and he needed an intense caregiving experience. He has made me feel more alive. Justin has helped me to feel that emotional connection and feel loving toward someone.

I do feel such loss, though. There is no way I can overcome what was taken by that experience. But I have a good life. I have security, I have a nice home, a beautiful new truck, plans to travel. So I made a life, but it's not the life I was entitled to or would have made if we didn't have this terrible episode that so brutalized all of us.

Donald Rander,

Army Sergeant, 1962–1983

Donald Rander was a twenty-three-year-old New Yorker when he was drafted in 1961. Although he didn't like it much, most of his friends back home were overdosing on drugs or getting killed and so he decided the military was his best shot at a good life, and he stayed on. He eventually got into military intelligence, and in 1967 volunteered to go to Vietnam because he believed it would help advance his career. In Vietnam he lived like a civilian in a villa about half a mile from the military post. Rander did security work, background investigations, and some classified work. But in 1968, toward the end of his assignment, he was captured by the North Vietnamese and spent the next five years and two months as a prisoner of war. About two thousand POWs are still missing.

In 1968 I was assigned to the Hue Field Office as the assistant special agent in charge, and my time was almost up. I was half packed to go home when Hue City was attacked on Tet. The North Vietnamese got between us and the military base in the city. We were cut off. There were two houses right next to each other and we all got together in one house; we decided to defend one place instead of two. There were nine of us. Two guys were killed in the first afternoon. We surrendered, because they had an overwhelming force and there was nothing else we could do. There were seven of us taken out of the house to another house down the street. It was kind of unreal to me. The Marines were in the city and the 101st Army was in close proximity. I had been wounded and sort of felt that we were going to be rescued, that the cavalry was going to come in and we would get out of there. It was so unreal. We were locked in close quarters, and I remember being very frightened that they were going to kill us. There was military in our group, and civilians who were working in Hue City for a construction company. There were about twenty-six of us captured in Hue City. But at the time when they captured me, my ID had been left behind and I wasn't wearing a uniform, so I was telling them I was a civilian, and they believed it. But I was worried—if the fighting got really hot, there was no way they could take us out of there. But eventually they did.

We were on a trail for a week or so in the mountains. It was a rocky, nonexistent trail through the jungle. We went through villages where the people wanted to kill us. They were yelling "Cut off his head" in Vietnamese, throwing stones, and spitting at us. At times our captors scurried us through the villages. It was obvious they didn't want us to be hurt, that they wanted to take prisoners.

I had several wounds, nothing really major. I had shrapnel wounds through my arm and legs oozing pus. My left ankle eventually became infected. One time we were crossing a river or a stream and we were jumping from rock to rock and on one of my jumps my leg gave out and I was just lying there. I wasn't about to jump up again, because my foot was really hurting. This one guy came over to me and pointed his AK-47 to my head and yelled something at me and I said, "Go ahead and shoot; I don't give a———" and I said a four-letter expletive. One of our guys came and helped me up and put me on his back and finished carrying me across the stream. It was after that they realized, by his act, that I wasn't faking it, that I was really hurt. So they allowed us to rest.

My ankle never turned gangrenous. But one day, I just went down to one of the rivers, which were reported to not be that clean because people defecated in them, and I said, "I'm going down to clean this thing." I just washed it out, prayed, and God heard me. I felt it was going to get gangrenous and I was going to be gone, but it seemed to heal from then on.

After stopping temporarily at two other camps, we were back on the trail and they took us to a camp near the DMZ* in North Vietnam. I was in solitary confinement, and they took me out to interrogate me. We were kept in a big hut with cells in it. Mine was three feet wide by five and a half feet long by six feet high; I could just stand up straight. It was made from lumber and we weren't allowed to talk. "Sparse rations" would be a kind thing to say. They gave us two meals a day, if you could call them that. We were given very little meat, if any.

At first I played my game. I was a civilian and we were involved in running checks on Vietnamese who wanted to work for the Army. I remember when the interrogator said, "Okay, Sergeant Rander, now that you told me that, why don't you tell me the truth." I looked at him. I was shocked. I just shut up and gave name, rank, and serial number. He sent me back to my cell and told me to think about this. I decided there were some things I knew that I couldn't take the chance of them torturing out of me, so I talked. I made up stuff. I lied and I cried and I told them so much BS. I gave them the names of 1951 Dodgers, using infielders as officers and outfielders as NCOs. If they asked me to write it down I would take four days to write it down, but I never wrote a confession. I just ranted for days.

They thought there was more to me than I was presenting, because my officer in charge broke. They broke him. I don't blame him. They just broke him physically and mentally. There was nothing he could do, and I had to defend myself against him. They thought I had personal meetings with high-level people in the government—i.e., the CIA and the executive branch—and that was

*The Demilitarized Zone was the dividing line between North and South Vietnam.

absolutely untrue but he told them that. They tortured him. I don't have any animosity toward him. He wasn't responsible. They started asking about one classified project that I'd been working on and I said I didn't know anything about that, or wouldn't answer. That's when they started putting me on my feet at attention for hours, or making me kneel for hours. After many beatings, one day out of desperation I tried something else. I said, "Hey, look, I am just a Negro soldier. I don't know anything about that. They wouldn't let me in on any of their secrets; they wouldn't tell me that stuff. They had me doing all the dirty jobs. I had to hire people and fire people. I ran the motor people. I didn't even have the combination to the safe."

My interrogator looked at me. I was on my knees at the time, and he said, "Sit up."

I sat on a stool that was much lower than his, which was part of their thing. He gave me a cigarette and then he said, "You see, this is what Johnson's racist regime is. . . ." And he went through this whole thing and I was like, Uh huh, and nodding my head thinking, Now, buddy, I got you. From that day forward, I was no longer tortured. When I fed their propaganda back to them about being a second-rate citizen and the white boys do all the important stuff even though I was the ranking NCO in the office, they believed it and it worked.

Next, they took us to a camp we called Skid Row. It was formerly a Buddhist monastery, and I was in solitary confinement there. I was there a little over a year, about thirteen months, in solitary confinement. It was a concrete cell, about seven feet by seven feet, with huge arched ceilings. The bathroom was a pot or a pail, which we were allowed to empty twice a day. Even though it was a tropical country, it was the coldest I'd ever been. I was chilled to the bone. We were stripped of our clothes and wore pajamas. We had two sets. They gave us soap, if you could call it that. It was damn near pure lye. There was a door in our area and when they opened it I saw a courtyard that had some chickens, and I'd watch them. The cells were next to each other. When the guards came to say something, we would learn to communicate to each other. By talking to these non-English speaking guards, we were really talking to these guys down the way. So when the non-English-speaking guard came I would say, "I'm Donald Rander. I'm an Army E-6. How you doing today?" We were communicating clandestinely; by talking loud we could let the other guys know that they were not alone. That was important.

We made a deck of cards out of paper. They would take us out every two days or so and ask us questions. Normally, they'd leave us there to write something, and we'd steal paper. The reason we stole paper was because they gave us a cigarette ration and you could take a regular cigarette—if you could call it that, because there was all kinds of crap in that damn thing—unroll it, take tobacco from it, and roll it in this confiscated paper. You could get six smokes out of one cigarette, and that was another game, an escape mechanism, because it's like, I'm getting up on this guy. They gave me three cigarettes and I've got eighteen smokes. It was a game you played, because you had to do something with your mind.

These little mechanisms I found, although inane and asinine, were defense mechanisms that kept you going, playing, active.

I prayed all day. I dreamed of family, home, birthdays, Christmas, and things I wanted to do for my family and things I hadn't done for my family. I wondered if my mother and father would be alive when I got back. I wondered what they were thinking. I was praying, begging them to hold on . . . begging myself to hold on.

I know one guy who died. There was nothing physically wrong with him. He just lay up in the bed and died. He gave up. He put the blanket over his head and died.

I kept track of dates with little notches on the wall. It was important for me to remember my daughter's birthday, my wife's birthday, my mother's birthday, the date we got married, Christmas and New Year's. I'd think, Gee, if I were home what would I do for that day? What did we do last year, or the years before? Remember that time . . . It was important. It was a survival mechanism. I would be like, Today is Mom's birthday. If I were home I would take her to such and such restaurant because I know that's where she would like to go. What would I have bought her? Gee, she likes scarves. So I would picture a designer scarf or something, and that would take the whole day to think about all that. Then I'd think about what I did last year on her birthday and try to remember a funny incident the year before that.

I also had a ritual when emptying my pot. When I emptied my pot the first time, that's when I made my mark on the wall. That was always the ritual: Come back, make the mark. It was important to keep track of days, and it was kind of easy. During the week the radio on the camp would come on with exercises. On Sunday they would have a children's choir on the radio, so you always knew when Sunday was. Maybe I'd say a few extra prayers that day because I was supposed to be at church. Sometimes, since I was an altar boy, I would go through the whole mass in Latin in my head.

There was one Vietnamese guard I would love to see again. He didn't smoke but he gave us cigarette rations. He'd come by my cell and he would look at me and I could see something in his eyes. Though we couldn't talk more than two words together, I'd ask for a light in Vietnamese. He would give me his half a cigarette to light mine and then he would walk away, leaving me with his cigarette. He carried messages written to me or mine to another prisoner. He eventually disappeared, but I think I saw him again when we were taking the bus to the airport on my way back home. I think at the time I interpreted his look as "I wish I could go with you" and mine being "I wish you could come, too, because I would love to show you around." I could imagine what he's never had. He did those few favors for us that could have got him killed. He did them out of compassion.

I was the only black prisoner. It was evident at one point that they tried to separate me from my fellow prisoners. On April 6, 1968, I was coming into camp from the trail and they had us in a cave to protect us. One of the Vietnamese who was traveling south had a radio. All I could hear was the riots in D.C. Then two or three nights later, one English-speaking Vietnamese said, "Do you know that Johnson and the CIA killed Martin Luther King?" They were trying to drive a stake between me and my fellow Americans—i.e., white Americans. I said, "Oh really." You can't say much to a guy who is holding a gun. There were several other instances when I was sharing a cell with a white American and a Filipino and I would see how they would defer to the white prisoner even though he was younger. I'm not thin-skinned, but you know when someone is talking down to you, when someone is trying to put you in your place.

On January 27, 1973, the peace agreements were signed in Paris. The camp commander came and brought most of us out to a common meeting room and told us the war was over. This man was telling us that we were going home soon and we would be moved to Hanoi, but we were just sitting there just as stone-faced as you please. This was the Asian way: Do not show emotion. I remember that to this day. They gave us a drink of Vietnamese brandy and we went back to our cells and we did not break a crack in our face the whole time. We walked through the courtyard and got back into our cells. They locked the door. When the Vietnamese were gone, my cellmate and I jumped each other. We were so happy.

The next day they took us by jeep to Hanoi, to the Hanoi Hilton.* The section we were in initially was called Heartbreak Hotel.† This is where they brought the initial pilot prisoners. My group and some of the others who were captured in Hue were reunited. There were twenty-something of us that were captured in Hue and we all had contact with each other at the Heartbreak Hotel. There were four sections of camps and an isolation section.

I wrote a letter while I was there to my wife. I wrote to her, "If you have an Afro, get rid of that." I associated Afros with radical comments because of the propaganda the North Vietnamese gave

---

*An old French prison that became a POW prison during the Vietnam War and was nicknamed the Hanoi Hilton.

†The Heartbreak Hotel was the prisoners' nickname for the brutal camp.

us. I knew I was fighting for the freedom of expression, the freedom of speech, and everything else, but when an American soldier is in the line of danger you haul that shit in. I have no respect and never will for those war protesters. My compatriots, the men and women in the service, were doing what their American government asked us to do. As blacks we were treated like less than the soldiers we were supposed to be. We were doing what our government asked us to do. Then these assholes—people enjoying all of the freedoms that all Americans have—turned their backs on us.

They released us in increments. When I was released in March 27, 1973, they made us take off every stitch of clothing. We had to strip, and the guards were watching us. They gave us clothes, underwear, pants, jacket, and a bag. We got on a bus and went to this building near the airport where they fed us two cans of Spam. I got to eat more meat in those last two months than I did in five years. We went to the airport and, one by one, passed in front of the American officials. They loaded us on a C-141 [an Air Force passenger plane] and we flew out of there. I can remember asking the Air Force flight attendant to ask the pilot to announce when we were "feet wet," which meant we were over the water and out of Vietnam, and when the pilot did, we shook up that aircraft. People were screaming and hollering.

One flight attendant asked if I was Donald Rander and then she handed me the *Stars and Stripes* [the military newspaper]. What happened was I gave my address to some of the guys who came out earlier and asked them to send my wife, Andrea, roses. So starting about two months before I came out she was getting roses every two weeks from me and they wrote it up in the *Stars and Stripes*.

The next joyful moment was when they told us we were "dry"— coming over California. They took us to a military hospital, and that's where I met the family. I met Andrea car-side and then made a few remarks to the press. I then saw my mother; Page, my daughter; and Lisa, my stepdaughter. Page was about nineteen months when I left. When I came back in '73 she was seven years and one month.

Later on that evening my mother and Page were next door in an attached room. Andrea and I were sitting in our room with a family assistance officer and we were all talking. She had put Page to bed. We were sitting there talking and Page came in the room. She wanted to read me a story: *Nobody Listens to Andrew.* She crawled into my lap with the book. In 1967, when I went to Vietnam, she was just learning how to say, "I love you, Daddy." She would give her mother fits because we would send each other tapes back and forth and Page would want to stay up late talking on the tape recorder. She would say, "Talking, Daddy; talking, Daddy, Mom. I love you, Daddy. I love you, Daddy."

That was all Page could say. When I came back home that night, Page is in her nightclothes, with her book, on my lap. I was a prisoner for five and a half years and I never cried. I hurt; I hurt like hell but I never cried. That night, I cried.

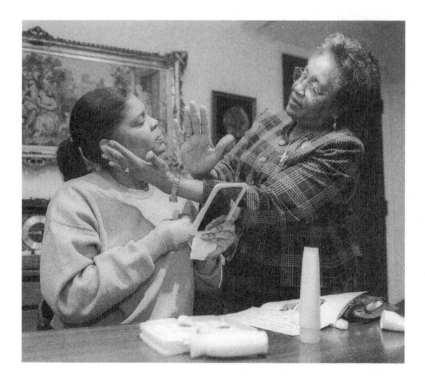

Olivia Theriot,

Air Force Lieutenant Colonel, 1959–1979

In the Air Force during the Vietnam War, flight nurses flew in cargo and commercial planes to base camps, where they picked up the wounded and transported them to medical centers in the Philippines, Thailand, and Okinawa in Japan. Nothing Olivia Theriot had done as a nurse prepared her for the brutal mutilations she saw on those planes. Tens of thousands of sick and wounded men were airlifted out of Vietnam. As doctors were rare, nurses shouldered the responsibility of caring for the injured. In the sky, they faced everything from hemorrhaging soldiers to those with missing limbs.

Theriot, of San Antonio, who now sells Mary Kay cosmetics, served two tours as a flight nurse in Vietnam. In 1968, the bloodiest year of the war, she flew into the Demilitarized Zone, which was the dividing line between South and North Vietnam, where some of the heaviest fighting of the war took place. Theriot, who was a captain, said the area smelled of death, but she never hesitated to make her flights. She knew she was needed in a way she had never been before and never would be again.

I was already a nurse living in Kansas City, and one day I was watching television, and a commercial advertising a career in the armed services came on. "Would you like to travel?" And I said, "Yes." "Do you want adventure?" I said, "Yes." "Do you want to further your education?" I said, "Yes." So I called all the branches of the service and tried to figure out which paid more. Once they had my name they all kept calling me. Finally, I decided to go. I wanted to go in the Navy because I loved the blue uniform, but when they started talking to me about sea duty, I said, "No way." So the choice was between the Air Force and Army, and I didn't like the color of the Army uniform, so I took my second choice, the Air Force. I eventually went overseas to France and was supposed to stay there only eighteen months, but instead I stayed four years. I loved everything about it. I learned to speak French. I loved the food. I loved the French people. I lived in a French village and the whole village adopted me. They felt sorry for me because I didn't have a man.

After France, I was assigned to the Whitman Air Force Base in Mississippi, and my chief nurse asked if any of us wanted to go to flight nursing school. She told me if I went, I wasn't going to get sent anywhere. It was more like education, so I went. As soon as I got out, I was sent to the Philippines Tactile 902nd Air Medical Squadron. I flew every day to Vietnam to pick up the wounded and bring people back to the Philippines. I flew in a C-118, which looked like a commercial plane, or a C-130 cargo plane. We dropped off supplies and then the plane would be reconfigured into something like a hospital. We would bring as many as one hundred wounded back; you could stack them five high. Usually we would have one other nurse and two or three medical evacuations technicians, who were kind of like nurses. We would bring them back to the Philippines or to Okinawa in Japan or Thailand. The patients had been triaged and a lot of them had definitive care.

But in 1968, the year of the Tet Offensive,* the care wasn't as good. All we could do was get them on the airplane and move them

*The largest offensive by the Vietcong in the Vietnam War, as they attacked many cities and towns in South Vietnam during the Vietnamese New Year, called Tet. Before the Vietcong were stopped, thousands were killed.

out as fast as possible. We flew two days straight without stopping, just back and forth from the battlefield. We were blessed: We got in and out. The men we picked up were freshly wounded. The whole ride we checked to see if they were in pain. We checked their IV fluids, checked their tubes. The plane was like a hospital with very small quarters and you had to make sure you had everything. They were not screaming in pain, but you knew when they were in it.

The first time I flew to Vietnam I was afraid. I had never been in a battlefield before, but afterward it was just like getting up and going to a hospital ward job. There were a lot of amputees, tracheotomies, hemorrhages, and battlefield injuries. You always looked at them and prayed that you could give them good care until you got them to the hospital. I was one of the few blessed nurses, because I never had an in-flight death. You didn't worry about eating or sleeping. All that mattered is your patients and getting them back to their families. The hours were long. I would get up at 4 A.M. By the time I got back it was 6, 7, 8, 9 P.M. Then I'd go back the next day. I flew every day until the end of my tour.

I talked to the guys about anything they wanted to know. This one fellow, he had both legs gone and one arm gone. He said, "Ma'am, this is the third time you brought me out of here." He was really concerned, but not about himself. He was from Hawaii and had a young wife back home whom he loved to death. They loved to dance. He knew he couldn't dance with her anymore. That was all he was concerned about. I tell you, they were wonderful men. There was such a need. You looked in their eyes and you knew they had been through a lot. You didn't get to know your patients as well, but sometimes it got distressing because they were so young and there were so many of them. But it was my job and I did what I had to do. I did my duty.

The Air Force let me organize a hot Christmas dinner for them on the plane. I got people at the dining hall to cook hams, turkeys, plum pudding, and mince pies, and we brought the food with us from the Philippines. I used my savings to get presents and gift-wrapped them. Everyone on the Air Evac team helped. We got flowers like hibiscus, bougainvillea, and poinsettia and put them all over the plane. Then we tied red stockings filled with nuts and candy to the seats and poles. We put stars and gold and silver angels on the ceiling of the plane. You would have thought we were giving them a million dollars each. It was something they didn't expect. They were so happy. We had Christmas for fifty-four patients and I got fifty-five thank-you letters. A corporal with a shattered jaw wrote two thank-you letters.

To me Vietnam was the epitome of my nursing career. Patients always need you, but there was a more special need for these young men.

When we had the dedication of the Women's Memorial in Washington, D.C., on Veterans Day in 1993, I was in the parade marching in my full dress uniform and this man ran up to me and just grabbed me and squeezed me.

He said, "Major, Major, You brought me back. You don't remember me but I remember you. You brought me out of Vietnam."

I hugged him back and he swung me around. That was such an emotional day for me.

I really loved the Air Force. It gave me opportunities I never would have had and in the end, I wound up really liking the uniform.

James Brantley,

Army Radioman, Specialist E-3,
1965–1967, Vietnam

ome historians call Vietnam the working-class war because so many who fought and died were from poor and working-class backgrounds. Young men could avoid the draft if they were in college or well connected, but that wasn't the case for many African American men. Blacks were not well represented on draft boards. In 1966, blacks accounted for about 1 percent of all draft board members. Seven states had no black representation at all.

When James Brantley got drafted, he had no way out of it. Brantley's mother died of complications minutes after he was born. When he was nine, he became a ward of the state and was shipped from foster home to foster home. At eighteen, he knew he wanted to be an artist, but he only lasted a semester at the prestigious Pennsylvania Academy of Fine Arts in Philadelphia because he didn't have enough money to pay for the second semester and no family to help him. After dropping out at nineteen, he was drafted, sent to Vietnam, and served in Saigon.

When I first got to Saigon one of the first things I saw was this toothpaste advertisement prominently placed. It was called "Light Bright" or something and they had this little caricature of a black

man, something you would see in the South, the Sambo thing. This guy is smiling with big white teeth, and that impacted me. I knew how they saw blacks based on that ad. In South Vietnam you could hear the young kids calling us "nigger." Someone else taught them that word. It wasn't a word from their culture, yet it seemed to pop up at times. I was in a bar one night and heard a Vietnamese citizen scream out "nigger" to a black GI. It was chilling, upsetting. An argument broke out after that between the two men. I had to try and stay calm, but it was hard. That ugly word was what white GIs brought to the Vietnamese people. It was not their hateful language—it was ours.

I remember sitting in a restaurant with my buddies, and we are black and white. The waiter comes up to me because I'm the darkest one there. He has a towel and he starts wiping my hand with this towel and I'm wondering, What the hell is he doing, and he keeps wiping. He then says, "It doesn't come off." I said, "Hell no, it doesn't come off, but a part of you might if you don't get off me."

We fought the war there, but then we were also fighting the war there that came from here.

Entertainment was segregated. They had the white bars and the black bars. Sometimes there would be a little mix. The lines were clearly drawn between Caucasians and blacks. At that time in the 1960s, we were very polarized, and that continued in Vietnam. I had some friends who were Italian, Irish, and Jewish, but for the most part we were segregated. If you walked into a cafeteria you would sit on the side where the guys were the same color as you, just like you would back home.

My supervisor was black but my commanding officer was white. I remember trying to get a promotion and having a really tough time. After you were in the service for a while, a promotion was mandatory, but they had a hard time giving it to me even though I had an excellent record. I watched all these other white guys get promoted around me but I had to wait longer than the mandatory time. It made it so I could never really feel comfortable.

. . .

A lot of the GIs both black and white were doing drugs like opium, heroin, and marijuana. It was a way to escape the craziness and uncertainty of Vietnam. The GIs would just go downtown and buy it or go and visit an opium den. But what was abused the most was alcohol. A lot of GIs had drinking problems.

They say God protects babies and fools. I don't know which I was. I guess I was a baby *and* a fool, because although I was sent to Vietnam I only got close to having to fight for my life a few times. One time I was in an Army bus driving through downtown Saigon. The driver was a friend of mine who was white and just the two of us were going to town, and suddenly he hit this young lady that was crossing the street on a bicycle. I remember seeing her hit the glass and the impact. She just flew off onto this main street where there were literally thousands of people around. So she's on the street sprawled out. I didn't know if she was dead or alive. We get out of the bus, but before we can get to her, a crowd surrounds us and they are saying something in Vietnamese we don't understand. And of course there's the ringleader, who is getting closer and closer to us in a threatening manner. We don't have any weapons and we are basically back to back, ready to defend ourselves, however futile that would have been. We were like Tony Curtis and Sidney Poitier in *The Defiant Ones* in downtown Saigon trying to survive. Then the police officers came over swinging these long batons, hitting people. They are beating them back and telling us to get back on the bus. We wanted to know what happened to the young girl, but they don't care about that. I guess they didn't want some kind of international story there. So we get back on the bus and back to our base. We never knew what happened. No story was ever printed about what happened to that young lady.

That wasn't the only time I wondered if I would get out of Vietnam alive. We lived in a rice mill that was converted into barracks. We got a letter one day from the local Vietcong and it said, "We are going to eat our dinner in your compound." We took that very seriously. The Vietcong were everywhere. They wore these black pajamas but they were homogenous with the entire population. That night they came across the canal toward our compound. They shot automatic weapons into our base. We were under attack. I was scared. I thought we were going to be overrun by these guys. We called for support and the helicopters came in and pushed them back.

I often saw some of the guys who did fight all the time when they came to Saigon for leave. They would come into town with weapons blazing. They'd cut off the ears of the Vietcong, make necklaces out of them, and wear them around their necks, like medals. They were out of control, crazy for what they had seen, what they had done, and no one was trying to control or help them. One time I was on the back of a truck with a few of them and they were just shooting off their weapons and throwing matches at Buddhist monks. The monks had been protesting the war by setting themselves on fire, so the "grunts" were making fun of them. Watching all that was just nasty, and there was absolutely nothing you could do to try and stop them.

When I came back to the States, I kissed the ground. Before you leave, you are talked to and they try and get you to reenlist. They offer you a bonus and guaranteed promotion. When I'm getting talked to, I'm looking at my watch and yawning, scratching my head. Finally the officer said, "You don't want to hear all this." And I said, "No." So he signed the paper and I was out of there. Finally, my career as an artist could resume. I came back to Philadelphia and the Academy on the GI Bill.

Vietnam impacted my life in ways I'm still trying to figure out. Most of today's politicians didn't go. Most of the wealthy didn't go. It was a grass-roots thing. Those who went had the least to gain.

Marie Rodgers,

Colonel, Army Nurse Corps, 1952–1978,
Korea and Vietnam

Marie Rodgers was born and raised in Alabama when Jim Crow segregation laws made it hard for any black woman to get ahead. But Rogers always wanted a life outside of Alabama. When she joined the Army Nurse Corps all her dreams came true.

Rodgers, now of El Paso, Texas, where she volunteers in the pharmacy at a local VA Hospital, made it to Paris and to many other cities in the world, but she also spent a harrowing year in Vietnam. She was a surgical nurse and held the rank of major when she joined up with the 24th Evacuation Hospital at Long Binh. A very busy hospital, the 24th Evacuation treated 9,010 patients in its first year of operation; over 5,900 received surgical care and 3,000 received medical care. About 3,782 were returned to active duty. The 24th Evacuation Hospital received its first Meritorious Unit Citation in the year 1967. For part of that year, Rodgers ran the operating room.

I asked to go to Vietnam because I had never been in combat. I had been an operating room nurse, since I served in Puerto Rico during the Korean War. I was fast on my feet; I had friends who were nurses in World War II and they were always talking about it. I wanted to know how I would perform. I wanted to know how I would do if I had to make do and be innovative.

In Vietnam we were right there. We did cranium, neurological, and facial injuries. We had specialists there. We had eight or nine head wounds a day. Every day we had lots of cases, lots of wounded. What we would do is irrigate and clean the wound, make sure the wounds would not get gangrene. Three days later, if there was no infection, we'd close them up, and then they were on their way home.

There was a big field where the choppers would come in. Most of the wounded were treated in the field by a basic medic who would take care of the guys who would fall in the field. The medic would have morphine, start an IV, and do some first aid. When he was finished the wounded would go on the chopper and they'd bring them to us. Then the guys with stretchers would bring them into pre-op, which had twenty to thirty beds. Each one of them was a table. Surgeons would be there to evaluate them. They would decide which ones were worse and who could wait. Surgeons made the decision. We were next door. They'd tell us how many guys they had for surgeries and I would get my staff of nurses ready to go.

Most of them were conscious when we were in the operating room. Some were scared but most of them thought they were in good hands. Some of them were surprised that the Army had a hospital, but this is what the Army promises the soldier: Wherever the soldier goes, there will be a hospital to take care of them.

I was glad we were there to help them. I have had people ask me why we [the United States] were in Vietnam. I never wondered why I was there: I was there to take care of the soldiers. As an Army nurse, there was never any doubt. To see the look on their faces, they were happy we were there.

Some wounds, I'd never seen anything like it before. One late afternoon a chopper came and brought in this patient. I can remember what table he was on. He had tourniquets on both legs. They were inflatable tourniquets, so they were tight. We put his legs on a stand on the table and let the air out of the tourniquets and then took them off both his legs. Even the surgeons were surprised. When you saw his leg from the knee down . . . There's a big bone, a tibia, and then muscles around your leg. When we took off the tourniquet, all you could see was his whole bone and strips of things hanging down like muscle and tissue. It was like someone had taken a knife and sliced him. The bone was intact but everything was hanging in little pieces, all the muscles and tissues. I don't know if he stepped on a mine or what. There was nothing to wash, just bone. I just stood there for a minute and said to myself, Oh, my God. Then the surgeon got busy and cleaned it up, dressed it. We knew we wouldn't see him again. There were major arteries there in the leg, and he had none. If you didn't get blood there, the leg would die. He had none, and lost so much tissue. No one thought he would live. He died that night.

A young soldier came in with a head injury when we were really busy. We put him flat on the table and draped his head. His legs were injured, too. He was a tall blond fellow, could have been a football player. He was awake and I kept saying, "Soldier," and then told him what I was going to do to him. He'd say, "Yes ma'am." I shaved his head and I talked to him the whole time. I told him he was going to be all right and we'd take care of him. He kept saying, "Yes ma'am."

He died on the table. We were short-staffed, so my sergeant and I had to take care of his body. We took him off the table and to an area in the back where we kept the bodies. I had to put him in a body bag. He's the only one I ever did that to. I had to put three tags on him—one tag on his toe, one tag around his wrist, and another on the outside of the bag. It was hard to get him in the bag. We started at the feet. It was like putting a snowsuit on a small child.

So we did this. Things were going through my mind, but I couldn't start moaning and groaning and all that foolish stuff. This is a young man who is dead and you have to do things right, but I kept thinking his momma doesn't even know he's dead. Somewhere in the States this boy's momma is doing something—going to work, cooking dinner, cleaning the house—and her boy is dead and she won't know until tomorrow.

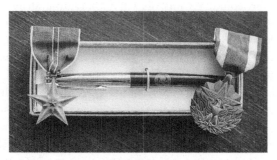

Marie Rodgers's medals hang on a box containing one of the pens used by President Johnson to sign an act to remove restrictions on the careers of female officers in the armed forces.

I got a Bronze Star* for what I did in Vietnam. President Lyndon Johnson pinned it on me. I got it because of how smooth my unit ran. In all these missions, the operating room supervisor, the nurses, we set everything up. All the doctors had to do was operate.

I was surprised I got the Bronze Star. I had no idea. My supervisor called me up and asked me to make sure my uniform was all right, but she didn't tell me why. She told me to go to the surgeon general's office at nine A.M. When I got there she and my chief nurse in Vietnam were there, and they told me.

A car came and drove us to the White House. There were all these generals, and all these people. I didn't know what to do. I was really nervous. Then the chief nurse read the citation and the president pinned it on me. He smiled at me and was real fatherly-like,

---

*Awarded to individuals who, while serving in any capacity with the armed forces in a combat theater, distinguish themselves through heroism, outstanding achievement, or meritorious service not involving aerial flight.

and after that we went to another room and would you believe, we drank wine at ten A.M.

My chief nurse gave me the rest of the day off. The ceremony was going to be on the news that night, so I called my daddy and told him I was going to be on television. He was so proud. He told everyone in Birmingham. Southern blacks wanted so much for their children, and they were stepped on all the way.

The Army system of promotion really helped. In other situations, as a black nurse, I wouldn't have gotten the kind of jobs I had. In the Army they always had to give you the job you were trained for, and with that, the rank. No matter what they thought of me, they had to put me in the operating room. My sister was in telephone repair; they would hire a white man and she'd teach him everything she knew and next thing you know, he would be her boss.

I can't say I experienced any racism. I was pretty fortunate, but there were not a lot of black nurses. Most times I was the only black nurse. Sometimes when there's a crowd of blacks, they are more apt to be prejudiced than when there is just one or two of us. I never worked with a black doctor. I never had a black surgeon work with me. I never even had a black nurse on my staff. I think I was blessed; I guess competent, too. I was always able to get the job done.

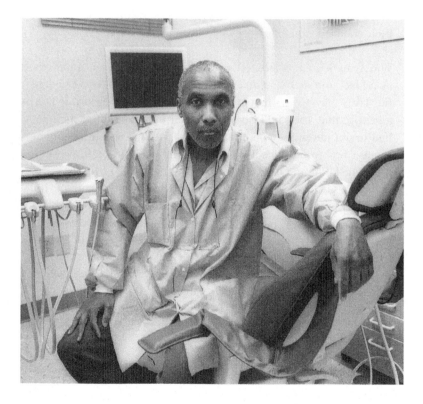

Alfredo Alexander,

Army Private, 1967–1969

 lfredo Alexander was a "grunt," an infantryman who climbed hills and crawled on his belly searching for an enemy he often found. In underground caves and tunnels, behind trees and bushes or in villages, Alexander's only job was to seek out and kill the enemy.

In 1966, black soldiers composed more than 22 percent of Army troops killed in action. The majority of black Vietnam-era servicemen, like Alexander, were in the infantry.

I'm from Panama, and immigrated here in 1960. I wasn't a great student. I got into college and flunked out. I really wanted to go back but I didn't have enough money. I became an American citizen to qualify for financial aid and wound up getting drafted while trying to get back into college.

I went in February of 1968 during the Tet Offensive.* When I got to Vietnam I was sent to the area from Chu Li to the DMZ† zone. We

---

*The largest offensive by the Vietcong in the Vietnam War, as they attacked many cities and towns in South Vietnam during the Vietnamese New Year, called Tet. Before the Vietcong were stopped, thousands were killed.

† The Demilitarized Zone was the dividing line between North and South Vietnam.

were a unit that was a light weapons infantry. There were a lot of black guys there, and our only job was search and destroy. You could spend the whole day going the equivalent of two blocks, because we were chopping thick bush and you never took a known trail because it was filled with booby traps. We would look for "gooks" and "dinks"—that's what we would call them. I don't know where the word "gooks" comes from but "dinks" comes from a term in Vietnam, which means "crazy." We'd see them running and we'd shoot back and forth. Sometimes we'd get ambushed.

The company had about ninety guys. We had no bases. We were just out in the jungle on the front lines. We wore plain old camou-flage green and it was not neat dress. I hardly wore a shirt. I wore mostly a T-shirt with bullets around my chest and a steel pot.* Heli-copters would come and drop boxes of C Ration meals. Let's say there would be a can of tuna, biscuits, cake, and mixed fruits. There are no bathrooms or baths in war. You don't brush your teeth; you didn't care about it. We all smoked marijuana. It was the only way you could deal. Guys would say you can find it along the trail; it was probably growing there but I was too naive to spot it. I never had any of my own—guys would give it to me. After so many months you get a furlough and are able to go to Bangkok and have the girls there. That's where they got the marijuana. They would get bushels of it. Even some of the officers were smoking it. No one was going to say anything. It was bad enough on the days you weren't getting shot at. You smoked a joint to get through that stuff, to find a brighter side. You're not thinking rationally. You're twenty years old. It made it so you could face the day.

We would also take rubbing alcohol, a little water, Kool-Aid—which our families would send to us—and mix it up. We had the Kool-Aid because we were drinking water from dirty streams. We would put pills in the water to kill the germs, and put Kool-Aid in the water to make it taste good. The helicopters would bring our mail to the jungle and take our letters back.

*Military helmet.

. . .

We would go out and do sweeps. In the middle of nowhere, sometimes you would come upon a village. You usually wouldn't see any
young men. You would see kids, women, old men and women. We
would secure the area, set up a perimeter. We would communicate
somewhat with the villagers. Then we would sit around there eating our C Rations and the kids would come over and look at us. I'd
give them a cracker or something, which was the biggest thing in
the world for them.

We never had problems with them during the day. The problems happened at night. We had a situation in this village—the kids
came over and everything was cool, but that night all hell broke
loose. We had set up a perimeter, so they knew where we were, and
they opened fire on us. They threw everything they had at us and
we fired back. The next day, when you looked at the bodies, they
were kids and young girls with AK-47s, assault rifles, on them. That
happened once or twice. I remember seeing the same kid, a little
boy I was talking to, whom I asked for some water and he asked
me for one of my biscuits, lying there dead with an AK-47 on him.

With that you lose sympathy. Some of the guys were cruel to the
people in the villages because they always suspected them of being
Vietcong. The outcome was that both black and white, but mainly
white guys, used to feed on that cruelty. I didn't do it, though.
They'd go to the village and say, where's the Vietcong and so forth,
and then they'd rough them up. There was this thing that we would
put on to keep the mosquitoes off but it would burn and guys would
spray it on the villagers to burn them. There was one white guy
from Alabama who used to revel in killing, not necessarily Vietcong but Vietnamese people. I remember him telling me that he
shot a girl, and the way he was telling me about it was really cold. I
would not shoot a little girl. It was a difference in mentality. For
him, killing was easy. For me it was not that easy. But most of the
black guys weren't like that soldier from Alabama. Most of us just
wanted to survive.

. . .

On this really nice day we were down in a valley and it was our day to get supplies. We had a hot cooked meal. As soon as we finished, the company commander said, "Saddle up," and we started hearing gunfire and getting hit. Really, we were ambushed. They were waiting for us, because our plan was to come up to this place where they were for the night. But they started to shoot mortars down at us. Now, my platoon was the lead platoon going up the hill. It was our job not to retreat. We had to crawl up that hill shooting. It was about six P.M. When we got up there the mortar was just coming down on us. They had us pinned down. We were returning fire the best we could. We weren't backing off, but at the same time I couldn't even look down. You would hear your buddy over there but you didn't know if he were dead or living or whatever. Only time you knew someone was alive was from hearing the sounds of the shots being fired. Everyone was staying low. I had got to the point that it didn't really bother me to be shot at, but the mortars were like hell. The mortars go up and they come down, and if you happened to be where it came down it would blow you to smithereens.

There was this one guy who was a medic. He said he heard that there were some guys who were hurt. All I remember is I moved over and he crawled up next to me and it was like two seconds later a mortar came in and it blew a hole right through him and that same mortar, the shrapnel, is what hit me. I remember lying there and a helicopter coming. The next thing I remember was they shocked my heart in the hospital. I must have died and they bought me back. From there I went to several hospitals. My arm was practically off. The wounds went up to my neck. I still have some shrapnel in me. I was in Vietnam for five months.

When I came back, at first I was just happy to be back. I'd sit in the park and look at the birds. I didn't want anybody to know. There were people who thought you were a baby killer or stupid for going in at all. One day I realized I needed to start getting myself together. So I took a ride up to a college to see if I could get in and start all over again. I told the dean I was a Vietnam veteran and all I wanted was to get my life together. I told her I wasn't a good student when I left but that I had matured, and she admitted me. I did well, and look at me now: I'm a dentist, even though I'm considered 40 percent disabled.

It's like therapeutic to talk about it. There a lot of guys out there like me who have made it. When people think of Vietnam vets, they think of angry people who can't get themselves together. Granted there are a lot of guys like that, but there is another side. Every story is different. Not all of us walked the same block. But people are not really open to listen to it. So I had to let it go—what's the point? But one thing that's always bothered me is that black soldiers' stories never get told. Whenever I turn on the television or see a movie about the war you don't see the black perspective, what we went through, and yet there were so many of us there. We got shot. Some got killed. Black families' lives were destroyed by the war, but when you look at movies we are either not there or just the backdrop. Vietnam wasn't like that. We were all there, all colors, doing our best to survive.

PART FOUR

Persian Gulf War

Dariton Battle,

Marine Sergeant, 1975–1995

The largest movement of Marines since World War II assembled in the desert after the Iraqi invasion of Kuwait. Between August 1990 and January 1991, approximately twenty-four infantry battalions and forty squadrons—a total of more than ninety-two thousand Marines—were deployed as part of Operation Desert Shield and Operation Desert Storm. But the main attack was on February 24, when the 1st and 2nd Marine Divisions stormed into Kuwait. Dariton Battle, a hardened Marine veteran, was the gunnery sergeant in the 2nd Division and was smack in the middle of the four-day ground war.

The Marines weren't my first choice. My first choice was the Air Force, but I couldn't get in. Once in the Marines, we trained in Parris Island in South Carolina. There were no kind words after we got off that bus for basic training. It was a shock, and I guess that was the purpose. And it worked. There was a lot of yelling, screaming, hollering, pushing, and a whole lot of words I don't say these days. They trained us how to walk, talk, eat, and dress, and everything was taught with an iron hand. Believe me, there was nothing polite about it. Being from Baltimore, the trend was everyone had to get a gold tooth, and I went there with that tooth and my name became "Goldie." It wasn't a good thing. Still, I worked hard and wound up graduating as a squad leader.

I was assigned to an infantry unit, the 6th Marines, based in North Carolina. We would go out with a Navy ship for about six months, traveling through the Mediterranean and Africa. We went to places like Kenya, Cairo, Egypt, and Spain. We would stop in various countries and do some training exercises.

I was in the infantry my first three years and then was transferred out for couple of years. I went to New Jersey and then back to North Carolina and was assigned to the 6th Marines in 1980, the time of the Iran hostages. We were out in the waters on an amphibious assault ship and were ready to go in and get them, but the call was never made. The hostages were released. I was transferred to the 2nd Marines, where I was the company gunny* for the 2nd Battalion. The role of the gunny is the best rank in the Marine Corps; anyone in the Marines will tell you that. The gunny is the logistics guy. You provide beans, Band-Aids, bullets, and more than that. If it doesn't work you make sure it gets fixed. I also was in charge of discipline. We went to cold-weather training and then to Okinawa, Japan, in 1990 for more training. We came back and three months later everything started going crazy over in Kuwait. I wound up going to Saudi with an advance party. We went in October of 1990. We prepared the site and contacted all the logistical areas, so when the main unit reached Saudi Arabia, everything was set up for them.

Three weeks later the rest of the unit, 296 Marines, got there. We were sleeping in cots under big huge tents. We stayed in that situation for a couple of weeks before we were told what our mission was. We had to pack up and move north. Every place we made camp, we had to dig in. You find it's tough digging into clay. Our little shovels weren't working too good. It was sand on top and clay underneath. When you think of the desert you think of hot, but when we were there, it was cold and it rained and we were miserable.

---

*Gunnery sergeant.

We were in a hostile area. We stayed in blackout situation, which meant we had to keep our vehicles' headlights off. We always set up a perimeter with two-men fighting holes. Anyone entering our perimeter had to know the password.

We weren't worried about firefights. Not too many armies can touch us in a firefight. But the reports we were getting said the Iraqis were going to use chemical agents. They wanted us to take these pills to protect us against chemical warfare, but I didn't take them because it was not known that the pills would help us. I had no clue about them. No one I knew, myself included, became ill as a result of being there.

We trained day in and day out. I wasn't scared at all. In the infantry, all the life you have is the infantry. There is no such thing as calling in sick or leaving work early to pay a bill. You have a schedule for everything. We trained for situations like Kuwait or Panama.* We were there so long, we were like, Come on, let's get this over with so we can go home. We knew what we were doing. My thought was, I'm not dying for my country, let that other guy die for his.

We moved five, six, seven times, and every time we moved we had to dig in. Sometimes you woke up the next morning in water. We didn't have tents, because we had to maintain a low profile. So we used ponchos to cover ourselves up and put over our hole, and that was enough to keep the rain out, sometimes. We didn't use our sleeping bags, which were warm but not comfortable at all. It was tough to maintain hygiene. The further away we were from the main campsite, the harder it was. We had to ration water. Our meals were MRE—Meals Ready to Eat.† Hot meals were provided maybe once a day.

*In 1989, the United States invaded Panama in order to bring its corrupt president, Manuel Noriega, to justice for drug trafficking. He is currently serving a forty-year sentence.

†Fully cooked food placed in high-strength pouches, with about a ten-year shelf life. A full MRE typically includes a main entree, a side dish, a dessert, crackers, and a spread.

The whole purpose was to fake out Iraq. Iraq thought we were coming from the sea; they never expected us to come from where we did. They set up the entire defense from the sea, and we had ships out there, but we were actually inching up, just like in checkers. It took us all of three months to get where we were supposed to. The whole time we were in a defensive and an alert position.

Then came February 24th. At 0500 we were sitting at the line of departure. Anything forward of that line was the enemy. There were minefields in front of us which engineers had to clear. We started moving at 0500. We proceeded forward and we were taking in some mortar rounds, but basically it was easy. There were a couple of Iraqis. These guys must have been left out there forever. They were in bad shape. They were hungry and their food was very low. They couldn't have known we were coming. The resistance was nothing. We took a couple of rounds and a few trucks were hit. But they had the casualties. Those who wanted to see Allah died. Then they started to give up. So many were coming to us, there were too many. We just kept going and told them someone in the back would pick them up.

First night, we had six prisoners in our camp—too many. We were not prepared. We were still in blackout mode. The only thing we could do to hold them at night was to put lights on them and tie their hands behind their backs and have them on their knees with lights on them so we could keep an eye on them. That night we thought we were going to get hit with chemicals. We had gas masks on, big heavy suits called MOPP—Mission-Oriented Protective Posture [suits to protect against chemical and biological attack]. They are terrible. It's hard trying to maneuver in them. Next morning we found more Iraqis buried under the ground, in these cavelike rooms underneath the sand that they had built. Once we realized they were there, they came out and gave up. They wanted to give up. They had gotten word that we were killing everybody, so they were too scared to come out that night. They wanted to make sure we could see their hands.

.   .   .

As we pushed toward Kuwait, we received a disturbing call: Iraqi tanks were heading our way and they were part of Hussein's prestigious guard, his great army. We were concerned about that. But then came the Tiger Brigade [2nd Armored Division], and they intercepted and wiped out the Iraqis. It was no match. It took us three days to get to what they called the "Highway of Death"; it was a four-lane road that led into Kuwait and it was the highway the Iraqis were using to get away. But the Cobras [Marine attack helicopters] and Apaches [Army anti-armor attack helicopters] tore them up.

When we got to the Highway of Death, there was a vehicle heading into Kuwait and we were told everything in front of us was the enemy. So our weapons company, our heavy guns, opened fire and hit the back of a vehicle. It swerved ahead of us and landed in the ditch. It turned out to be a news reporter and a cameraman. Fortunately, they weren't hurt, but the reporter was so upset. She said we had no business shooting at her; we had to explain she had no business on that road. We stopped at the Highway of Death, and that was pretty much it for us.

We lost a lot of guys, not by enemy fire, but accidents. One guy had a hand grenade on his belt clip and he pulled the clip by accident and it killed six or seven guys. Guys had to dig holes to sleep in, and they put scrap metal on top of the holes. The metal was so heavy the holes caved in and the guys suffocated. We had situations where we were shooting at each other because of poor communication. We had marking on vehicles so the planes would know who we were; well, one or two didn't have the marks and got knocked out. Most of our tragedy happened not by the enemy but by our own mistakes.

Overall it was a quick war. The waiting was the worst part of it. Going from October through May living in dirt, rain, windstorm, cold, and heat was hard. But you know the first rule of combat is if you take the head the body will fall. We wanted to go on. We wanted to go on to Iraq. We wanted to take out Hussein's guard.

There was no reason for us to stop. We should have kept going. We should have finished it.

When the war was over, people headed back to the States, but we had to wait until our number was called. I'm a sports nut and started thinking of competitions. Whoever had the best-looking hole would win a five-gallon jug of water to take a shower. One guy, he made a room. He made four steps, dug benches, made a table and a bed, all out of dirt.

We had a spade tournament. The winner got a five-gallon jug of water and his own tray pack [a hot meal]. Talk about some trash talking. It was great, though.

We left Kuwait in the latter part of May, and the welcome we received was awesome. It was great. We flew in on an advance party, landed in Cherry Point, North Carolina, and there wasn't a lot of hurrah there, because we were just a small party. But when we got back to Camp Lejeune in North Carolina, all our wives were there and they had a banner along the highway welcoming us home. People were waving flags. The band was struck up. At the base all the flags were up and it was the best thing to see your family once you got off the bus. It was like something like you would see on the main drag of New York. It was a great feeling.

Even though I'm no longer a Marine, I never really left. I still work on a Marine base in Virginia. I work at a Marine hotel doing the same sort of thing I did as a gunny. I don't have one regret about the time I served. Not many people can say that about a job they did for twenty years.

Janet Pennick,

Army, 1974–1977;
Army Reserves, 1977–1998

anet Pennick was a forty-four-year-old mother, Army reservist, and deputy sheriff in the Philadelphia Sheriff's Department when she was sent to the Persian Gulf War. During her six-month tour she endured endless Iraqi missile attacks on her base and lived in daily fear for her life.

As a member of the 304th Civil Affairs Unit, she was part of a team whose duties included coordinating laundry, shower, and mail facilities with the Saudis and arranging ramp space for coalition aircraft and warehouse space for everything from medical supplies to captured weapons. She was one of about 41,000 women who served in the war, most in support roles. About half of those women were African American.

When the problems in the Gulf started, I had just been promoted to first sergeant in the 304th Civil Affairs Unit, an Army Reserve unit that oversees everything. We control the comings and goings of the troops; we give them their assignments, take care of paperwork and all the little details that make things go smoothly. I wasn't very confident that I could do the job. My first sergeant had volunteered to go to the Persian Gulf and I was asked to take his place. At first,

I told them no—I didn't have the experience or knowledge to do the job—but another officer pulled me aside and said he would help me. He was a really nice man who knew a lot and I thought, Well, this could be my chance to move up and learn. So I took the job. Next thing you know he gets sent to the Gulf. Shortly after that they decide to call us up.

When I first heard I was getting called up, I was in shock. I had been watching the war on TV. I knew they were calling up troops, but I didn't really believe they would call my unit up. When I heard they were sending us, I was so scared and nervous. I didn't think I knew enough to do the job right. But I spent so much time calming down other people, it kept me focused. So many wives came to me crying because they didn't want us to take their husbands. I spent more time counseling than anything else. My own daughter was away at her first year of college, so I didn't have to worry about her.

After three weeks of training at Fort Bragg, North Carolina, we were sent to Saudi Arabia. We were attacked as soon as we landed on the airstrip in Saudi Arabia. As we got off the plane, there were Scuds* overhead. I had heard of Scuds. We had been trained to deal with them, but it was nothing like the real thing. We had just landed and all of these troops are all over the airfield and we are trying to get our troops together, and all of a sudden something like a huge gigantic firecracker exploded in the air. No one really knew what happened. There were sirens going off, but no one knew what they meant. Then suddenly these officers come running out of the building yelling and screaming, "That's a Scud. You are under attack. Come into this building."

We all had on our M-16, gas masks around our waist, and our MOPP suits on and everyone started running toward this airport

---

*Erratic and dangerous missiles, with the potential to carry chemical and biological agents. They were first deployed by the Soviets in the mid-1960s. The missile was originally designed to carry a 100-kiloton nuclear warhead or a 2,000-pound conventional warhead, with ranges from 100 to 180 miles. The Scud's principal threat was the potential of its warhead to hold chemical or biological agents.

building in a panic. You're supposed to keep all your gear with you, but everyone was dropping their gas masks and M-16s on the ground. People were running scared; you could see the fright on their faces. I was running, too, just as fast as everyone else. Running for my life. My heart is pounding. It was hot. When I looked up I saw like sparks and flashes of light and debris falling down. Our missiles were hitting the Scuds. But it dawned on me as a I was running toward the building that I'm the first sergeant and I better make sure my troops all get to safety. So I went back to make sure everyone had come to the building. When we got in the building we were told by the officers that when you are under attack you need to put on your MOPP gear and have a better system, because we were a mess. This was my welcome to Saudi Arabia.

After leaving the airfield we drove for miles in the desert. Everywhere you looked was desert. Camels roamed everywhere, like stray dogs. If you opened your mouth, camel hair and sand would fly in. When we got to our base, the first thing we had to do was tape the windows with an X; in case shrapnel hit, the glass would stay together and not shatter all over the place.

They told us when we first got there not to worry, that they don't normally fire Scuds during the day, only at night. Well, we believed them. That first night nothing happened, so our hearts stopped racing. But after that night it started again. The Scud missile attacks never stopped; they were fired day and night. They would last all night, or just a couple of hours. Sometimes I spent an entire night sitting on the floor of the hallway or other "safe areas." Other times we were there for two hours and were told "all clear." I'd get back in bed and the sirens would go off again. We never knew if there were chemicals in the Scuds.

I was scared. We were scared. Many of us had diarrhea for weeks because we were so scared. I had to keep my fears inside. I couldn't let the troops know. I had to perform. After a while I adjusted to it. Just as if you live in a neighborhood where there's drugs all around you—you adjust.

After six months it was time to leave. I was ready. I was anxious.

I couldn't wait to get out of there. All of us in my unit got back in one piece, but I know of other civil affairs units that had casualties. As soon as I got back to Philadelphia I transferred to the school unit, which was a training unit that would not send any of its troops to the Gulf or any other war zone.

I'm proud of what I did there—I just wouldn't want to do it again. When I got back, troops were being diverted to Somalia and other places. I didn't want to go. Everything about war was horrible. I couldn't wait to get out of there.

For twenty-one years I served in the Army, and I had a lot of good years. It's just war that I don't like.

J. Alexander Martin,

Navy E-4, 1989–1991

I n the Gulf War the Navy's main goals were to get control of the Persian Gulf, protect friendly ships, stop Iraqi trade, and launch an attack against Iraq. Much of the heavy armor, supplies, and troops of Desert Shield and Desert Storm were taken to the Middle East by the Navy.

On one of these ships was smooth, handsome J. Alexander Martin of Queens, New York, a young man with a flair for fashion design but no outlet for it. Martin's ship bought Marines to the Middle East in the beginning of the war. Although he saw no action in the Gulf War, it was during his time in the Navy, when he stood on the deck surrounded by endless ocean, that he figured out how he was going to make it big back home.

After leaving the Navy he cofounded FUBU, a Manhattan-based international, multimillion-dollar sportswear company aimed at the hip-hop generation.

I grew up with two sisters and a brother. My brother and I were adopted. I didn't have any goals or aspirations. I was just there, living life. I come from a place where sitting on the corner was the biggest thing that we did all day. I'd talk about what I was going to do, when I was going to do it, but it was totally unfulfilling. I never went anywhere; my parents wouldn't let me. My friends would go on tour with L.L. Cool J, who was our childhood friend, but I couldn't go. My parents weren't really strict, they just wanted me to walk straight. I always stayed to myself and wondered what I was going to do, how I was going to make a lot of money. For years I sat in the dark and thought and thought and thought. In junior high school I started working in a clothing store. My father always dressed really sharp and so did my sisters, so I watched them. I don't know if fashion was a knack for me or I just picked it up.

By the time high school ended, I wasn't doing very good. I was always on punishment because I didn't do well in school. So my father said, "It's time for you to leave, and I strongly suggest you go into the military." The Navy picked me. I really wanted to go in the Air Force but I took a general test and got assigned to the Navy. I didn't really want to go. I wanted to go to fashion school. In high school I used to change my clothes two, three times a day and all I cared about was clothes, clothes, clothes. When I went in, I asked them if they had anything in fashion and they said no, but there's other things.

At home I was suppressed. In the Navy I was somebody. I was doing what I wanted to do, when I wanted to do it. I grew as a person. I blossomed. I saw the world. Before that I was a kid living in Queens, and that's all that I saw; I may have gone to Brooklyn twice, Manhattan three times.

After I finished training school in Orlando, Florida, they sent me on deployment. I went to Okinawa on a carrier that deployed Marines. I went to Japan, the Philippines, and Mexico. The best time I had was in Singapore. I shared a hotel room with my friend. Just to be able to walk in another country is so exciting, you can't even put it in words. I thought to myself, If the guys from the corner could see me now.

We were out to sea a lot of time. We would work and work. I went to school for radio and teletype repair, but I never got to fix anything on the ship because I was just starting out. There was a lot of repetition of things. When I was there I never understood why I was doing the same thing over and over. In the beginning, all I ever did was clean up. I had to clean the ship, and paint the ship, over and over again. What they were doing was trying to get me familiar with the ship.

About 25 percent of our ship was black. The guys used to say my superior officer, who was black, wasn't racist toward white people but toward blacks, because he was so hard on us. He was always on our backs, trying to get us to work harder and be perfect. He treated us that way because some of the white officers looked down on us. He had to work hard to get where he was and he wanted us to work hard and be the best.

While I was on tour, things were getting beefed up in the Gulf. My ship deployed Marines. We were in the vicinity, so we had to steam through there. We started practicing what would we be doing if we were on full alert. You had to go to the station you were assigned to. At one point I was driving the boat. Then I became a lookout.

When we got to the Gulf we deployed the Marines and never got too close to the action. I was in the Gulf for about two months. We were the first ones there; I was there at the beginning. The Navy was your life, you were almost brainwashed—you didn't think about the danger, the fact that there was a war going on.

I got really worldly. I've been to Korea and all these places through the Navy. By the time I was twenty-something, I'd been to seven different countries. The military gives you a different perspective. It can lead you the right way. It taught me attention to detail, which I didn't have in my life. They make you do the same thing eight hundred times.

While on leave in Queens I got in a car accident, hurt my back, and left the Navy. I was living in D.C. and couldn't find a job. I lost my apartment. My parents told me to come home, and I went to work for Macy's and then I quit. I said, "Next time I come back here I'm going to sell to you." Then I came across these guys I knew forever we were friends since junior high school. One of the guys was making hats and I was like, Hats? What do you mean hats? I wanted to be in fashion, so I convinced him and the other guys to get into fashion. I said we could make clothes and a whole bunch of other things. I put the money I got from the insurance payment into the business, and look at where we are now.

I'm glad I went to the military. As far as fashion and life, it broadened me. I'm well rounded as a person. Now I can go uptown, downtown, midtown, and still deal. I can put on a suit and put on regular clothes and hang out on the corner with the best of them. The Navy wasn't as complicated as the world I am in now. The Navy was a beautiful experience. It got me out into the world, it got my mind going. It's about learning and growing and being your own person. It happened. It's done. I never talk about it. When we give interviews, it's my partners who talk about it. I'm no "Land of the free, home of the brave" type waving a flag in front of my house, but God bless America.

Lester Outterbridge,

Army Private, 1968–1970;
National Guard Specialist 4, 1988–1996,
Vietnam and Persian Gulf War

Lester Outterbridge was thirty-seven when he joined the National Guard to make extra money for his family. Two years after that he found himself in the middle of the Gulf War, one of 53,000 Guardsmen. Sent initially to Saudi Arabia, Outterbridge was a dispatcher in a unit of truckers who ran the military bus system.

Since leaving the Gulf, Outterbridge has battled Gulf War Syndrome, a mix of illnesses including heart, circulatory, throat, and stomach problems, a severe skin rash all over his body, and post-traumatic stress. Too ill to work, he runs a small support group for veterans who suffer from the syndrome, and is an associate pastor of a local church.

I'll never forget when the air war started in the Persian Gulf. The young boys were blasting their rap music and we heard a loud explosion. One of my sergeants screamed, "Mask," and I don't know how I didn't have a heart attack. First thing I remembered to do was to stop breathing, put on my mask, and blow out like they

showed us. I tightened my mask and prayed that I had it on right. There wasn't even time to put the rest of my stuff on before we had gotten hit. The explosion was so loud. No one was telling us nothing and so we put on our whole MOPP gear, which was a protective suit that had masks, rubber gloves, rubber boots, pants, and a jacket. Guys were freaking out. I'm claustrophobic, and I remember it was so hot because you had on your regular uniform, plus the MOPP suit on top of that. They told us if even a little bit of any chemicals or gas got on you, it could kill you.

No one was hit when the Scud hit our building, but just a quarter mile down the road at another building, twenty-five guys were killed in the same Scud attack. Sometimes there would be two and three attacks a day, mostly early in the morning. When guys were sitting around playing cards, I would practice putting on my gear. Everywhere I went I always took my weapon and my gas mask. I didn't care if I went to the bathroom or if they asked me to go next door, I took everything with me.

My commanders told us we would just be dropping off troops and food to other bases around the desert. Then they switched it around on us and said we are actually going 480 miles over to Iraq on a secret mission. They wanted us on the ground there. They wanted us to fight.

I couldn't even tell my wife, Gloria—that's how secret the mission was. We were supposed to hit the Iraqi Republican Guard. I've never been in combat. I'm a family man, and this is no Hollywood thing, this is the real thing. I was close to forty and I had been through a lot in my life and had overcome a lot of things, but I was scared.

Before we pulled out I held up this sign and had my picture taken. It said: "Live from Iraq to my beautiful wife Gloria and my children and grandchildren and brothers and sisters, loved ones and dear friends. I love you all. Lester." I carried the picture with me, and if something happened to me, I wanted the picture to be given to her so she would know my final thoughts.

Lucky for us the Marines, and the tanks up front, did their job. They were kicking butt up there, so when we got to the outskirts of Kuwait, you could just see these Iraqi tanks all burned and blown up. There was just complete devastation. We spent about a month at a base in Iraq near a place where the Allies stored up all the captured ammo, and then they detonated it. This big black cloud of smoke came out of it. It was like a black wall that was about three hundred and some miles long, and every troop in that area was affected by it. It covered everything—our uniforms, our tents—and since the war was over, nobody said to put on our MOPP gear. It only lasted a few minutes but afterward everyone was coughing, sneezing, and scratching their eyes. It was so dirty, it was like having chimney soot on. It was horrible.

We left in May, and as soon as I got back home, I started getting chest pains. I had pains in my chest and arms over there, but the military doctors would just check my blood pressure and say everything was okay. They did tell me I had post-traumatic stress and anxiety disorder. They wanted to keep me in the hospital but all I wanted to do was be with my family. So they discharged me. I felt sick over there but it didn't really bother me to the great extent that it did until I came home.

The trauma would come up in horrible ways. Once in the middle of the night I woke up and I found myself choking Gloria, the love of my life, and I didn't even know it. Sometimes, I would get up in the middle of the night and *Bam!* I'd punch my fist through a wall. I'd wake up screaming. I can't even sleep with Gloria anymore, because I don't trust myself; I don't know what could happen.

I also have skin problems, dermatitis. I have esophagus reflux problems that are very bad, and bad muscle pain. I have heart problems. My blood levels are always elevated. I have rotating pains, so I'm in pain all the time. They know I'm sick but they don't know exactly why. I have about thirty visits to the emergency room in a year. Simple things, like cutting the grass, I can't do anymore. If I lift something heavy, I feel like my arms are on fire. I used to love to go to the mall and be around crowds, but now I can't take it; I sit in the car and wait while my family shops. Even now when I hear a siren for a volunteer fire company, in my mind I'm back in Iraq. At first the Army told me it was all in my head but then I heard of more veterans getting sick. They later admitted we were exposed to some bad gases, but I had to get my senator and congressman to help me get social security benefits.

All veterans ask for is to be treated with dignity, whether you are in a combat zone or not. We should be treated with honor and respect. That's why I'm an advocate now. There are too many African American soldiers that are suffering. You see a lot of black men out here who are in their late forties, fifties, even sixties living on the street. You know why? Because they gave up. They gave up on the American system because they think, I went over there, I came back messed up, and nobody wanted me.

We Gulf War vets were welcomed home. But in the same breath we were called liars. We were told that there was no gas, even when it was proved that gas was used. Why are so many of us getting sick? Why are so many of us getting Lou Gehrig's disease and brain tumors? One lady in my unit died of a brain tumor. I have a friend in a wheelchair. I can't tell you how many guys call me for support. I've had guys say they are ready to kill themselves. The only thing that keeps my sanity is my belief in God.

PART FIVE

War on Terror

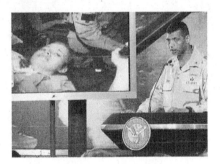

Anthony LaSure,

Air Force Captain, 1989–1999;
Air National Guard Lieutenant Colonel, 1999–2004,
Operation Noble Eagle

The 177th Fighter Wing, stationed at Atlantic City International
Airport in New Jersey, was preparing for routine bombing
training exercises on the morning of September 11 when two
passenger airplanes, manned by terrorists, crashed into the Twin Towers
of the World Trade Center. For the next eight months, Maj. Anthony
LaSure, a copilot for American Airlines, flew his F-16 fighter plane in
Operation Noble Eagle, flying above New York, Philadelphia, Washington, D.C., and Camp David and prepared to shoot down any commercial
plane that appeared suspicious. A father of four, he would have done
whatever necessary, without hesitation, to protect America.

I grew up in Atlanta and didn't know anything about aviation. In
the ninth grade I decided to go into the Air Force Academy. I
wanted to do some intelligence work, that kind of thing, but they
said you have to go into pilot training. I could have pulled some
strings to get out of it but my grandfather said, "Hey, it's kind of
hard to get into that pilot training thing. Why don't you give it a
shot. If you don't like it you can quit, but it would be a shame to

miss the opportunity." Those words changed my life. Look at me now—I love it and I can't think of anything else I would rather do.

I've gone to the Gulf twice—once while on active duty and once in the Air National Guard. Both times were for Operation Southern Watch. That's the big operation that happened after the Iraqis started bombing the Kurds in the nineties. After we had already beat Saddam up, he was bombing his own people. They've got some serious ethnic problems over there. So, under the United Nations Peacekeeping Force, we flew over Iraqi space. I did that twice. The first time I wasn't sent to fly over the no-fly zone per se— it was to drop bombs. Saddam had said, If you are going to spy on me I'm going to shoot you down.

I was a brand-new wingman, and I was there to drop iron. We got short notice. We flew nonstop from Moody Air Force Base in Georgia to Bahrain, seventeen hours' flying time. It was hell. It was just me, alone, in my fighter plane. They gave us the option to take "goat pills." They are like speed. It's a diet pill that helps you stay awake, because there's no place to land and take a break. I didn't take mine because I was scared the whole way across, and so was most everyone else. That's how I stayed awake. When you travel like that you don't fly high like airliners. You fly low because you have to stay close to the refueling plane, a KC-10, which you hit over and over again. In higher altitudes we wouldn't have enough power to keep up with them. Sometimes the weather got so thick and all of a sudden I'd lose the big KC-10. One minute it was just three feet away from me and then it was gone. It was like that the whole ride—scary. When we landed I was so messed up, the crew chief had to help me out of the airplane.

I left the Air Force and joined the Air National Guard, because it's not conducive to being around your kids very often. Family was the big driver for me getting out. I also got a job as a pilot for American Airlines and joined the Guard. Being in the Guard I get to have my cake and eat it, too. The Guard is part-time.

On September 11, I was at the Guard, scheduled to fly to Fort Drum, New York, to pick up bombs and do some live exercises. I remember it like it was yesterday. It was this bright, clear morning—it was crystal clear. I thought, This is going to be such a nice day to fly. We were all on the runway taxiing and all of a sudden I got this call to come back in. Right away I think, Oh, my God, something's happened to somebody's kid, because this has never happened before. So we all came back and everyone was nervous, everyone has the same idea. The chief came up to us and said a plane just rammed into the World Trade Center. We went inside and watched the constant replaying of it on television. Our commander was running and scrambling around, trying to figure out what was going on. We got no word on what to do but to their credit our guys loaded bombs and missiles right away. Then we saw the second plane crash and we got permission to take off, heading toward New York.

I gave a briefing to my wingmen before we went up and we talked about how we were going to take these commercial planes down if we had to. I said an airplane won't fly without a tail, so we were going to do zigzag patterns and saw the tail off with our bullets. There was a thought to hit the engines, but a plane would fly fine with just one engine. If we hit the fuel cells the plane would also still fly for a while.

It was the scariest thing. The Northeast corridor is the busiest airspace in the country, probably the world. You have Philly, Newark, La Guardia, Kennedy, and all these little airports in between. It's hard to get a word in edgewise as far as traffic control goes, and the procedures are tight. But we took off that day and our Guard fighter planes could fly any altitude we wanted to at any airspeed. My wingman and I flew fourteen hours that day; it was a long day. We were intercepting airplanes and everyone was told to land. We were going off after stragglers, and then we heard about the flight over Pennsylvania: It crashed while we were heading out that way. They kept us over the water to intercept boats, because they didn't know what was going on with that either. No one knew what was going on.

While I was flying, I compartmentalized. I didn't have time to think about the attacks or the lives that were lost. I actually saw the towers fall; we were airborne when they fell. Wow, that was something else. All you could see was the smoke. It was a dust pile—that's all we could see. I had a hundred messages on my cell phone. My wife said our phone was ringing nonstop. I didn't even have time to tell her I was all right. I work for American Airlines, and word was out that it was an American airplane and a United airplane that crashed. My friends got worried. I knew the flight attendants onboard those airplanes. After we came down, we had to have a scotch and say, What the hell is going on? It was the closest feeling I ever had as a man of being violated. The whole country felt violated and confused.

For three months straight I only had only twelve hours off, a day to go home, see my family, and sleep. This had never been done before—patrolling our own country. So it took them a while to figure out the command structure and get organized. Then I started flying over Philly, New York, D.C., and Camp David. We ran on adrenaline and flew virtually nonstop. After eight months it got mindless. There was no more adrenaline. My friends asked if I would have shot down a plane that had my friends onboard. If I had to do it, I would have done it. I would have taken those few lives in exchange for the thousands who would have died if a commercial airplane hit a target. The plane is like a guided bomb full of gas. It's just a silver tube. I don't see the people's faces. I was trained to do this, to deal with the worst situations, put it behind me and take decisive action.

I was the only African American who flew in Operation Noble Eagle, the patrolling of the skies after September 11. I don't even think we make 1 percent of the military pilots. I literally know every black pilot—there are just a handful. It's a hard gig to get into. It's like golf, tennis, hockey—it's money, and a hard nut to crack. You just don't walk in. I'm in charge of the hiring for the Guard, so I get to see what civilian guys have to go through. If you don't have a private pilot's license, don't even apply. You have to do good on an SAT-like test. Guys and girls have to have a score of 95 or better just to get an interview. I've been doing it for two years and have only seen one black guy apply. It's a shame. There were more black fighter pilots in the 1940s than there have been since. The military is not that high on our social scale anymore.

As for my white colleagues, their dads are pilots. They are second- and third-generation pilots. They eat and sleep the stuff. When I went in I didn't know anything about it. The number of blacks going into pilot training is narrowing down. I have no idea who is watching it or who cares.

War is not what it was in Vietnam, when they were sending blacks out to die. It's white boys out there dying, because they are the frontline fighters. Why are all the blacks in the rear? Why aren't they at the tip of the spear where all the glory is? I have no idea what they are going to do about it, but the military is not attractive to our people. How many times have you been in church and heard a preacher say, "Go join the Army"? If anything, you hear him say, "Stay out of the war."

Eric Mitchell,

Air Force Captain 1996–present,
Operation Enduring Freedom,
Operation Iraqi Freedom

fter the horror of September 11, twenty-seven-year-old Air Force Captain Eric Mitchell had his first assignment as a flight commander. A pilot in the 32nd Air Refueling Squadron, based at McGuire Air Force Base in New Jersey, he was to fly his KC-10 Extender, a plane that carries gas for fighter planes, over Afghanistan. Mitchell was apprehensive, but like many Americans wanted justice for the loss of so many innocent lives. During this interview, Mitchell was once again on edge. He was on call to be sent to the Gulf as the United States prepared to go to war with Iraq.

My parents were in the military. They met in the military. My father was a loadmaster on C-141 cargo planes. While I was growing up, he was always going on trips around the world, and bringing stuff back. I was born on a base in Delaware, and then we moved to South Carolina, then Alabama, and then I moved to Buffalo, New York, and lived with my grandmother. Ever since I can remember, I always wanted to fly—at whatever cost, I always wanted to fly. I would ride my bike out and just sit at the end of the runway and watch the planes and think, Wow. It was always a dream, a goal that I worked toward, and I was blessed to have it happen for me.

I graduated from high school and got an appointment to the Air Force Academy in Colorado Springs. First, in 1991, I had to go to the Air Force Academy Prep School for a year because my SAT scores weren't high enough to get into the Academy. I started the Air Force Academy the next year and struggled there academically. I went to high school in the inner city of Buffalo, and there was no way that I was prepared for the type of academics I encountered at the Academy, even though I went to prep school. I would be up until two, three in the morning studying. I graduated in 1996 and got a slot for flight school, which I finished in 1997, and was number one in my class.

When most people think of Air Force pilots they think of fighter pilots, but I didn't want to drop bombs and blow up stuff. I was an anomaly in that sense. I had no desire to fly fighter planes. I wanted to fly medevac. I wanted to save people, not kill them. I had to get "counseled" for not wanting to fly fighters. My first assignment was in Germany, flying medevac. I would fly patients in Europe to the military hospital in Germany. That was the most incredible three years of my life. The impact I had on people, directly contributing to their health and well-being, was the most incredible feeling. But then it was over and I was sent to McGuire Air Force Base and started flying KC-10 Extenders, which is basically a flying gas station. Now I'm working as part of an air refueling squadron, doing things that directly support dropping bombs. It has been a big adjustment for me. Before, I always went to the back of the plane and little kids drew thank-you pictures. That's what I miss the most. My job now is not about people but about the machine. So I got stationed at McGuire Air Base in New Jersey in December of 2000, went through training, and shortly thereafter I became a flight commander.

My first mission as a commander was over Afghanistan. I left in November, and twelve hours later we landed at the base we were stationed at in the Middle East. The reason the KC-10 Extenders are such a valuable asset is because we can take gas and we can stay in the air forever, because we can give away all our gas and then another tanker can give us some more. A fighter loaded with a whole bunch of bombs can't fly for a long time, and none of our bases were close enough that they could take off and make their targets. We took fighters to Afghanistan, refueling them, then they dropped the bombs. Then we took the fighters back, refueling them all the way. I had four planes on my wing, with twenty thousand pounds of bombs.

Our missions were long; the average mission was ten to twelve hours. There was a lot of concern, because it was the first time we were dealing with a country where our enemy was landlocked. In previous conflicts, in Vietnam and the Gulf War, the tankers weren't in harm's way—they were over the water. This was the first time in history we had to put the tankers right over the country we were dealing with. We had no defenses at all, so we were basically up there like sitting ducks. We knew exactly what was happening on the ground; we heard the explosions when the bombs dropped.

I was in charge of a crew of four. I was always uncomfortable. It was my responsibility to get them back to their families. I was always thinking, What if? I would look for a place to land in case something happened. Anytime we were over Afghanistan, I was uncomfortable. If you let yourself get comfortable, that's when you make a mistake. Sometimes they shined lights on us and would shoot. All we could do was turn, climb, and get away from them. The hardest part, though, was giving gas at night, with your lights off, because that's how we had to fly. One bad move and that was it. It was dangerous, but there's been only one fatal accident, and I was not involved in it.

The first time I saw explosions, people in my crew cheered. I thought that I would feel proud of that, but I wasn't. I was happy to be doing my part, but I thought with those explosions, people were dying, people probably my equal, a person just like me or not like me who is just trying to feed his family. Perhaps he's been taught to hate Americans. He doesn't, but has to say he does to feed his family. What choice does he have? I had thoughts like that at times.

. . .

Then there's my private life. I'm single. Last year, out of 365 days in the year, I was gone about 250. It's very hard to meet people. You're here for a while, then you go, come back, and are like, Hey, do you remember me? I'm resigned to the fact that marriage won't happen for me until I get out of the military. Guys leave their kid when they are two weeks old; they come back and try to hold them and the kid screams. The first time I came back from Afghanistan, I had kind of an empty feeling. I looked outside and everyone was outside waiting with cameras. There were wives, girlfriends, and kids. Then you have the single guys. We got out last. That was kind of emotional.

I never think of what I achieved. I'm thankful for the Tuskegee Airmen and Gen. Benjamin O. Davis.* Their stories were motivating for me. If it weren't for them, I wouldn't be sitting here now. I had the opportunities to pursue a dream, a goal. I was blessed with determination. I don't think I'm that smart at all—like I said, I struggled through the Academy. I was surrounded by a whole lot of people who were smarter than me, but none worked harder than me.

Two weeks after this interview, Mitchell was sent to the Persian Gulf to prepare for the upcoming Operation Iraqi Freedom. During the war he served as assistant director of operations, overseeing four hundred combat aerial refueling missions a day at the Al Dhafra Air Base in the United Arab Emirates, and was awarded two Oak Leaf Clusters for outstanding achievement during the war. His citation read, "The lynchpin of operations, Captain Mitchell's efforts enabled a swift and decisive dismemberment of Saddam Hussein's tyrannical Iraqi regime and the rebirth of freedom for the Iraqi people."

But Mitchell had a tough time emotionally during the war. During the bombing of Baghdad, he wrote, "I appreciate all the prayers and support, but I think we all need to focus on the Iraqi people and the immense terror they are experiencing. We need to focus on the families of those

---

*Commander of the 99th Pursuit Squad of the Tuskegee Airmen.

twenty-one individuals who have already passed on to a place where there is indeed eternal peace. We need to focus on that eighteen-year-old Marine who is on his way to downtown Baghdad, who only a year ago was enjoying his high school prom. We need to focus on that unborn child who will never get the opportunity to meet his father because he gave his life in a land so far away. We need to focus on the newlyweds who will never get the chance to enjoy their first wedding anniversary. We need to focus on the Iraqi soldier who will die because he was trying to find a way to feed his family. Pray 4 a quick ending to this."

He came home on April 27, 2003.

Vincent Brooks,

Army Brigadier General, 1980–present,
Operation Iraqi Freedom

Modern warfare is not fought only on the battlefield—public opinion and support is just as vital and strategic. When Americans turned on their television sets for news of the war in Iraq, it was Brig. Gen. Vincent Brooks's face they saw. General Brooks was a young commander on the rise, working with the Joint Chiefs of Staff, when he was tapped by the military's top leaders for the job.

His crucial role was to keep the media and the public up to date on what was going on in the battlefield. He became the assuring face and patient voice that the public turned to every day to know that this war was going as planned, and would be won.

My father was a general in the Army, but there was never any pressure on us to go in. In my case, I wanted to be a doctor for as long as I could remember. I used to find myself studying *Gray's Anatomy* and took foundation courses in premed. But in my late junior year of high school I took a greater interest in the military and started feeling like I wanted to be an Army doctor. Then I saw the transformation my brother went through after coming back from West Point: the confidence, the way he carried himself. It drew me in, and I went to West Point.

I'm still trying to figure out how I got my position in Iraq. I served with Gen. Tommy Franks in Korea from 1996 to 1998 in the 2nd Infantry. He was the division commander and I was one of his subordinate commanders. He helped me develop and placed a great deal of trust in me. He then asked for me to serve for him again when I got out of War College in 1999. He was at the 3rd Army and I became his chief of plans. Again, I felt he had a great deal of trust and confidence in me. I later found myself working with the Joint Chiefs of Staff in June of 2002, and we were in a prelude to a potential conflict with Iraq. By January it was clear that if we were going to approach war we had to use the lessons we'd learned from Desert Storm. We had to communicate what was going on to the rest of the world. It's strategic, not just public relations–related. Franks and the other leaders were looking for the right person to do this operation. They wanted someone who could articulate details of the operation, get it across to the public, and who wouldn't mind dealing with the media. It was an opportunity and an honor. It was not something I would have sought out, but someone I respected called me to duty.

It didn't escape me that I was a black man. I thought, Was it because I'm black or because I'm qualified? I've come to the realization that it doesn't matter. I'm black and will be black tomorrow. An opportunity was extended to me and I was qualified to do it.

I was stationed in the operation base in Qatar. It was like other places in the Middle East: very dry, everything is sand-colored, and there is little vegetation. We were on a base developed by Qatar and us. We were literally inside a warehouse that had expandable command-post modules. It was a good facility, a bit austere but very effective.

My days were long, but I always kept thinking that my old unit, the 3rd Infantry, were having endless days. They were on the front lines. They couldn't trust anything around them. Even a pregnant woman in a burka could explode a bus. They were caught up in dust storms and never knew if they would make it through the night. My roommate in West Point was killed, as were men I knew in my old unit. So no matter how long my days got or how tired I was or how much fortitude it took to stand in front of the world's media, nothing compared to what the troops were going through. That was always on my mind.

My job started every day around 4:30 A.M. I woke up, dressed, and then went to see what was going on. I would check the operation databases and look at electronic pictures of the units. We could see literally where the vessels were in the Red Sea and our units in Iraq. Then I'd go to the operation center and see if there were any key items I should be aware of. Next stop was to go to the media center and link up with the people who were providing me support. Our camera crews had pictures and video images I would use, such as those from the night when Jessica Lynch was rescued. We would look and see what images we could use to help us articulate what happened. Then there was another briefing of updates and discussing priorities. I would meet with General Frank and bring him up to speed and get his insights, which was very important because I needed to represent him. Then I would make notes and comments on my opening remarks. There was also a video teleconferencing meeting with commanders in other countries like Saudi Arabia, Turkey, Pakistan, and the entire region. Before I went in to the briefings, I knew the views of all the commanders. Right before we went on, we would meet with the planning team again and our media affairs folks would warm me up. They'd highlight the issue of the day and bring me up to speed on what was already shared so that we would be consistent and I could think my way through things and not go on cold. The daily briefing was 3 P.M. our time, 7 A.M. Eastern time. We would have the first word of the day and then subsequent briefings. After the briefing we would do question-and-answers.

Sometimes I would feel the same sort of tension you get before game time. I felt as though either I fumble for the nation or deliver. But I'd get past the nervousness and emotion. I knew I had to perform my duty. It was a mission, and I took it as seriously as being on the line. The hardest part was being removed from the action. You want to be with the troops. My old brigade was the one who took the airport outside Baghdad. I was so proud of them, and I missed that I did not have the chance to lead them.

After the briefing I would go into the green room and we would review how things went and what we would do differently. Then I went back to the media center and prepared for the 7 P.M. briefing. Afterward I'd eat a salad that was usually brought to my desk, get more updates, go through E-mail, and try to get out of there by 10:30 to 11 P.M.

I was somewhat unemotional about the end of the war—I long ago learned to subdue emotion. I felt cautiously optimistic. The fighting was over but there was still a great deal of danger out there. It's more complicated now. I knew that some of my friends were still likely to get shot and some of their wives would never know the story of their valor.

By April 24, 2003, my role had come to a terminal point. I came back to my position with the Joint Chiefs of Staff. I came home to my wife, Carol; we have been married for twenty years, and that was a good reunion. While I was away she had attended several funerals of friends who lost husbands in Iraq. I was glad to be home with her and my family and have been delighted with the reception. When I go out to public places there is a tremendous amount of support for what we did.

African Americans in the military set an example for our nation. We have to open doors and then with one hand, reach back for others. We have come a tremendous way in the military, but we have a long way to go. There are areas where blacks have not held command, not for any other reason than there is not enough density.

If there is an institution that is always looking to the future but also looks at its experiences in the past and tries to learn from it, it's the military. The military develops and seeks people who have talent. If you take talented people and extend opportunities to them, they will succeed. That's the lesson for our nation—extend opportunity to those who qualify, give an opportunity to excel.

The military has opened the door for a whole lot of people—from the humble enlisted man who as a result gained access to education, to the secretary of state, Colin Powell, who is one of my heroes. It's not finished yet. There are doors that still have to be opened. But African Americans have always been there, in every war, meeting every challenge, no matter how hard. We have always stood up to the call.

Acknowledgments

Yvonne and Ron

We'd like to thank our agent, Diana Finch, who believed from the start; Dawn Davis, our encouraging, supportive, and patient editor, who made this book a reality; and editorial assistant Darah Smith.

Big thanks to all those who helped us find veterans: Lt. Diane Weed of McGuire Air Force Base; Leigh Anne McNamara of the Women's Memorial; Jerome B. Milburn of the Montford Point Marine Association; Lt. Luz Colon of the 177th Fighter Wing of the Air National Guard; Rev. Leonard Smalls, Bob Hill, and Marti Trudeau Donahue.

We will forever be indebted to the veterans in this book and their families, who welcomed us into their homes and treated us so kindly. We thank you with all our hearts.

Yvonne

I want to thank the spirit of George Ingram and Doug Culbreth, a Vietnam veteran who got me to care more than I ever thought I could.

I could not have done this without Ron Tarver, an amazing photographer who somehow managed to shoot what is in my heart.

I was encouraged and supported by so many, but especially Jenice Armstrong, Karen E. Quinones Miller, Scott Flander, Jack Morrison, Marisol Bello, Nicole Kilcullen, Janet Barag, Grace Lawrence, and my mother, Ramona Latty.

Special thanks to Barbara Laker, an incredible and compassionate writer, for her time and her ear; Shelley Spector, who was with me step by step and her faith in me never wavered, even when mine did; and my two daughters, Nola and Margo, who inspire me every day.

Ron

The making of this book has been one of the most rewarding experiences in my photographic career.

It would not have been possible without the understanding and commitment of several people. I am eternally grateful to my partner and friend, Kristin Winch, for her support, and to our children, Wesley, Michaela, and Riley.

A sincere thank-you is in order for Clem Murray, director of photography at the *Philadelphia Inquirer,* and Berford Gammon, assignment editor at the *Inquirer,* for deciphering my constant requests for days off while working on this project. Thanks as well to Eric Mencher, Ron Cortes, April Saul, and Michael Wirtz for lending me their skillful eyes during editing.

Finally, my very special thanks to Yvonne Latty for asking me to join her in this journey. Her true professionalism, dedication, and excitement smoothed the few bumps in an otherwise very pleasant ride.

CPSIA information can be obtained at www.ICGtesting.com
Printed in the USA
LVOW04s0539070515

437530LV00015B/1/3/P

9 780060 751593